T0319406

Electromagnetic Metasurfaces

Electromagnetic Metasurfaces

Theory and Applications

Karim Achouri
École Polytechnique Fédérale de Lausanne
Lausanne, Switzerland

Christophe Caloz
KU Leuven
Leuven, Belgium

Registered Office
John Wiley & Sons, Inc., 111 River Street, Hoboken, NJ 07030, USA

Editorial Office
111 River Street, Hoboken, NJ 07030, USA

For details of our global editorial offices, customer services, and more information about Wiley products visit us at www.wiley.com.

Wiley also publishes its books in a variety of electronic formats and by print-on-demand. Some content that appears in standard print versions of this book may not be available in other formats.

Library of Congress Cataloging-in-Publication Data is applied for

9781119525165

Cover Design and Images by Mireille Clavien

Set in 9.5/12.5pt STIXTwoText by SPi Global, Chennai, India

Contents

Preface

Over the past decade, metasurfaces have been developed into a novel paradigm of modern science and technology. They offer possibilities to manipulate electromagnetic waves that were simply unthinkable even 10 years ago. They have already led to a myriad of new applications, reported in uncountable scientic publications. Moreover, they have recently started to make their way into the industry in a most promising fashion. Such a spectacular development has brought metasurfaces at the time of this writing to a point of maturity that is optimal for their introduction into the syllabus of all researchers and engineers active in the broad eld of electromagnetics across the microwave, terahertz, and optical regimes of the spectrum. This book, the rst textbook on the eld, responds to a pressing need for a unied document as a support to this syllabus, and opens horizons for further developments of the field.

What is the point about metasurfaces? After all, their visual appearance is essentially similar to that of frequency- or polarization-selective surfaces and spatial light modulators, which have been known and extensively used for several decades. The point is. . . DIVERSITY. Metasurfaces represent generalizations of these devices, which are respectively restricted to ltering or polarizing and phase-only or magnitude-only wave manipulations. Metasurfaces indeed offer a universal platform for unlimited control of the magnitude, the phase, and the polarization of electromagnetic waves to meet transformation specications of virtually unlimited complexity. Moreover, in contrast ix to their voluminal counterparts, they feature a low prole that implies easier fabrication, lower loss, and, perhaps surprisingly, even greater functionality.

This diversity would be of little practical interest if it were not accompanied by a proper instrument to master it. Fortunately, such an instrument has progressively emerged and crystalized as a combination of bianisotropic surface susceptibility models and generalized sheet transition conditions, the former properly describing metasurfaces, whose subwavelength thickness prohibits Fabry–Perot resonances, as a sheet of equivalent surface polarization current densities, and the latter representing a generalized version of the conventional boundary conditions, via the addition of these currents to the induced sources. This global modeling approach allows to systematically manipulate the enormous number of possible combinations of the 36 degrees of freedom of bianisotropic metasurfaces, both in the analysis and characterization of existing metasurfaces and in the synthesis and design of metasurfaces according to specications. It constitutes, therefore, the backbone of the book, providing, in addition to its design power, deep insight into the physics of metasurfaces and quick information on their fundamental properties.

The book constitutes a coherent, comprehensive, self-consistent, and pedagogical framework that covers the essential theoretical aspects and practical design strategies of electromagnetic metasurfaces and their applications. It is designed to provide a solid understanding and efficient mastery of the eld to both students and researchers alike. To achieve this goal, we have organized this book in a logical fashion, with each chapter building up on the concepts established by the previous ones. Considering the plethora of publications in the eld, we naturally had to select out what we considered as the most important and relevant developments, hoping that the presented material would be x sufficient to provide comfortable access to all the non-explicitly treated topics.

Chapter 1 provides a general denition of metasurfaces as well as a historical perspective of their development. Chapter 2 describes the fundamental electromagnetic media properties pertaining to metasurfaces. It presents the general bianisotropic constitutive relations, reviews the concepts of temporal and spatial dispersion, and recalls the Lorentz and Poynting theorems. Chapter 3 extends these concepts to the metasurfaces, establishes their modeling principles in terms of bianisotropic surface susceptibilities and generalized sheet transition conditions, and extracts from this model several fundamental physical properties and limitations. Chapter 4 applies these modeling principles to the synthesis of linear time-invariant, time-varying and nonlinear metasurfaces, and illustrates them with several examples. Chapter 5 overviews four analysis schemes that may be used to simulate and predict the scattering response of metasurfaces. One of these methods is based on the Fourier propagation technique; two are based on nite-difference methods, one in the frequency domain and the other in the time domain; and the last method is an integral equation method, which is particularly well suited for 3D computations. Chapter 6 details several fabrication metasurface technologies that pertain to the different parts of the electromagnetic spectrum. It shows how dielectric and metallic scattering particles may be used to realize metasurfaces at both optical and microwave frequencies and demonstrates several techniques to tune the shape of the scattering particles to achieve desired responses. Finally, Chapter 7 discusses three representative metasurface applications, involving respectively the properties of angle independence, perfect matching, and diffractionless refraction, and constituting representative examples of state-of-the-art metasurfaces.

1

Introduction

Metasurfaces are subwavelenghtly thick electromagnetic structures that may be used to control the scattering of electromagnetic waves. Because of their subwavelength thickness, they may be considered as the two-dimensional counterparts of volumetric metamaterials. However, compared to metamaterials, metasurfaces offer the advantages of being less bulky, less lossy, and easier to fabricate in both the microwave and optical regimes.

This chapter presents a general description of metasurfaces and metamaterials and the types of electromagnetic transformations that they can realize. Section 1.1 provides a historical perspective on the origin and concept of metamaterials, while Section 1.2 discusses the emergence of metasurfaces along with their applications and synthesis strategies.

1.1 Metamaterials

The origin of the term *metamaterial* can be traced back to a 1999 DARPA[1] Workshop [183], where it was introduced by Rodger M. Walser to describe artificial materials with electromagnetic properties *beyond* those conventionally found in nature [25, 29, 41, 165]. Obviously, the concept of artificial electromagnetic structure has existed long before the appearance of the term metamaterial. One of the most well-known and ancient examples of metamaterials is that of the Lycurgus cup, dating to the fourth century AD, which is a type of dichroic glass like those used in the conception of stained glass. Back then, the optical properties of glass were tuned by adding various types of metallic powders during its fabrication process but the lack of proper understanding of the interactions of light with matter meant that these processes were essentially made by trial and errors [145].

---

1 Defense Advanced Research Projects Agency.

Electromagnetic Metasurfaces: Theory and Applications, First Edition.
Karim Achouri and Christophe Caloz.
© 2021 The Institute of Electrical and Electronics Engineers, Inc.
Published 2021 by John Wiley & Sons, Inc.

It is only much later, in the second part of the nineteenth century, that people became properly equipped to study and control the propagation of electro-magnetic waves thanks to the newly established theory of electromagnetism of Maxwell [100]. Soon after, near the end of the nineteenth century, Bose introduced what was probably one of the first examples of artificial electromagnetic structures in the form of a polarization rotator made of twisted jute that was supposed to imitate the chiral response of some sugar solutions [20]. Similar types of chiral responses were also introduced by Lindman in the 1910s and 1920s using spiral-shaped resonating structures [94]. The concept of "artificial dielectrics" was then pioneered by Kock in the 1940s, which originally consisted in inserting metallic patches inside layered metal–dielectric media. Based on this principle, Kock realized the first-known artificial microwave delay lens [79]. During the same period, split-ring resonators were proposed by Schelkunoff and Friss as a mean of controlling the permeability of artificial materials [144].

Until the middle of the twentieth century, artificial electromagnetic structures were mostly used as a means of achieving desired values of permittivity, permeability, and chirality within reasonable ranges. It is only since then that more extreme material parameters started to be investigated. For instance, artificial materials with a refractive index less than unity were introduced by Brown in the 1950s [22] while, in the 1960s, Rotman associated the response of wire media to that of conventional plasma [141]. In 1968, Veselago mathematically described the response of materials exhibiting both negative permittivity and permeability and demonstrated that such material parameters would correspond to a negative index of refraction [170]. Later, during the 1980s and 1990s, important theoretical and practical developments were made toward the realization of bianisotropic media and microwave absorbers [165]. However, it is only after the beginning of the twentieth century that the field of metamaterials really started to attract massive attention. It is notably due to the first experimental demonstration by Shelby et al. [150] that a negative index of refraction was indeed feasible and the associated concept of the perfect lens imagined by Pendry [120]. The interest in metamaterials grew even more when, in 2006, Pendry and Smith proposed and realized the concept of electromagnetic cloaking, based on transformation optics [121, 146].

At the beginning of this section, we provided a very general and broad definition of metamaterials. To be more specific, we shall now define more rigorously what we mean by a metamaterial. Throughout this book, we will consider that a metamaterial is an artificial structure consisting of an arrangement, usually periodic, of scattering particles. The effective physical size of these scattering particles, as well as the distance between them, must be smaller than that of the wavelength of the electromagnetic wave interacting with them. This prevents the existence of diffraction orders or Bragg scattering due to the lattice period and allows the metamaterial structure to be homogenized into effective material parameters. Note that this condition implies that photonic crystals [71] and electromagnetic band-gap structures [174], which typically exhibit unit cell size larger than that of the wavelength, cannot be considered to be metamaterials since it is impossible to describe them in terms of effective material parameters. The effective material parameters of a typical metamaterial stem from the combined effects of the chemical composition, the shape, and the orientation of their scattering particles. From a general perspective, the scattering particles may be composed of metal or dielectric materials; they may be resonant or non-resonant structures; they may be active, lossy, nonlinear, and even time-varying or contain electronic circuitry.

1.2 Emergence of Metasurfaces

Although the interest in metamaterials was at its peak in the first decade of the twenty-first century, it is their two-dimensional counterparts – the metasurfaces – that progressively became the dominant source of attention. The main reason for this sudden rise in interest is explained by the fact that metasurfaces are easier to fabricate, less bulky, and less lossy than conventional volume metamaterials [48, 57, 58, 105].

While the interest in metasurfaces only became substantial in recent years, the conceptual idea of controlling the propagation of electromagnetic waves with thin artificial structures is more than 100 years old. One of the earliest examples of this concept was most likely proposed by Lamb, at the end of the twentieth century, who investigated the scattering properties of an array of metallic strips [87]. This was soon followed by the polarization reflectors made by Marconi in the 1910s. However, the most important progresses in this area were made in conjunction with the development of the radar technology during World War II and which continued through the 1950s and the 1960s leading to a rich diversity of new systems and technologies. We illustrate some of these systems in Figure 1.1. For instance, Figure 1.1a depicts a Fresnel zone plate reflector, typically used in radio transmitters [23], whose working principle is based on the concept of the Fresnel lens already pioneered 200 hundred years ago. Figure 1.1b illustrates another very well-known operation, which is that of the frequency-selective surfaces used as microwave frequency filters [110, 139]. Then, Figure 1.1c represents a rather modern version of a reflectarray antenna (or reflectarrays) [61, 98, 107], whose technology, dating back to 1970s, consists of an array of microstrip patches implementing the functionality of conventional parabolic reflectors. Note that the first types of reflectarray antennas were designed using short-ended waveguides instead of microstrip patches [17]. Finally, Figure 1.1d illustrates the working principle of a transmitarray, the transmission counterpart of a reflectarray, which appeared already in the early 1960s [75, 101, 160]. This kind of systems was

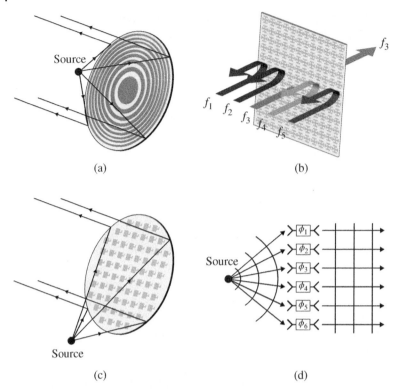

Figure 1.1 Examples of two-dimensional waves manipulating structures [2].
(a) Fresnel-zone plate reflector. (b) Reflectarray. (c) Interconnected-array lens.
(d) Frequency-selective surface. Due to their structural configuration, (a) and (c) do not
qualify as metasurfaces. On the other hand, the structures in (b) and (d) can be
homogenized into effective material parameters and hence correspond to metasurfaces.

typically made of an array of receiving and transmitting antennas connected
to each other by a delay line, whose phase shift could be tuned to shape the
wavefront of the transmitted wave. Later, in the 1990s, microstrip patches were
proposed as potential replacement for the interconnected antenna array used to
implement transmitarrays [34, 86, 132]. However, the transmission properties of
microstrip patches were then plagued with low efficiency as their phase shifting
capabilities could not reach the required 2π-phase range. It is only around
2010 that efficiency transmitarrays able to achieve a 2π-phase shift range were
demonstrated [44].

The structures and systems presented in Figure 1.1 are the precursors of current metasurfaces. However, they are not the only ones. Indeed, there are many
other examples of electrically thin structures that would be today considered as

metasurfaces. For instance, consider the case of near-field plates [51, 65, 66] and impedance surfaces converting surface waves into space waves [43, 97], which, unlike the structures in Figure 1.1 that are restricted to space-wave transformations, can accommodate both space waves and surface waves.

As of today, there are countless examples of metasurface concepts and applications that have been reported in the literature and it would therefore be impossible to provide a complete description of all of them. Instead, we shall now present a non-exhaustive list of some typical applications of metasurfaces. A very common application is that of polarization conversion [45, 60, 69, 90, 99, 157, 173, 180], where a linearly polarized wave may be transformed either into a circularly polarized one or be rotated using chirality [77]. Metasurfaces have also been commonly used to control the amplitude of the incident wave and, for instance, been used as perfect absorbers by altering the amplitude of the fields so that no reflection or transmission can occur [19, 40, 91, 135, 168, 172, 177]. Nevertheless, one of the most spectacular aspects of metasurfaces is their wavefront manipulating capabilities, which is typically based on the same physical concept used to realize Fresnel lenses [106] and blazed gratings [38, 112], and which have been used for applications such as generalized refraction, collimation, focalization, and holography [1, 12, 46, 47, 55, 62, 63, 95, 96, 136, 153, 154, 175, 178, 179, 181]. Finally, more exotic applications include, for instance, the generation of vortex beams characterized by orbital angular momentum [76, 93, 127, 161, 171, 176], stable tractor beams for nanoparticle manipulation [128], nonreciprocal scattering for electromagnetic isolation [33, 36, 137, 159], nonlinear interactions [89, 169, 182], analog computing [131, 156], and spatial filtering [117, 151, 152].

2

Electromagnetic Properties of Materials

In order to effectively model, synthesize, and analyze a metasurface, we need to understand how it interacts with electromagnetic waves. From a general perspective, understanding the interactions of a given structure with an electromagnetic wave requires two fundamental prerequisites: (i) a description of the structure in terms of electromagnetic material parameters and (ii) the availability of appropriate boundary conditions. The purpose of this chapter is to address the first prerequisite, while the second one will be dealt with in Chapter 3.

The concepts discussed in this chapter are presented for the general perspective of volume materials, and hence deal with 3D material parameters. However, they also essentially apply to metasurfaces, which are modeled by 2D material parameters throughout the book, as will be established in Chapter 3.

This chapter presents a general description of the medium parameters and constitutive relations of materials. It also provides a detailed discussion of the physical properties that are inherently related to these parameters. Specifically, Section 2.1 introduces the conventional bianisotropic constitutive relations. Section 2.2 describes the temporal response of matter (temporal dispersion) and the mechanism responsible for resonances. It also provides the fundamental relationship between causality, and the real and imaginary parts of material parameters through the Kramers–Kronig relations. Section 2.3 presents the spatially dispersive nature of matter (spatial dispersion), which helps understanding the fundamental origin of bianisotropy. Sections 2.4 and 2.5 derive the Lorentz reciprocity theorem and the Poynting theorem, respectively. Based on the Poynting theorem, Section 2.6 then deduces energy conservation relations for lossless-gainless systems both in terms of susceptibilities and scattering parameters. Finally, Section 2.7 classifies bianisotropic media according to their fundamental material properties.

Electromagnetic Metasurfaces: Theory and Applications, First Edition.
Karim Achouri and Christophe Caloz.
© 2021 The Institute of Electrical and Electronics Engineers, Inc.
Published 2021 by John Wiley & Sons, Inc.

2.1 Bianisotropic Constitutive Relations

The electromagnetic response of a linear medium may be expressed in the frequency domain, using the MKS system of units, as

$$\mathbf{D} = \epsilon_0 \mathbf{E} + \mathbf{P}, \tag{2.1a}$$

$$\mathbf{B} = \mu_0 \left(\mathbf{H} + \mathbf{M} \right), \tag{2.1b}$$

where $\mathbf{D}$ (C/m^2) and $\mathbf{B}$ (Wb/m^2) are, respectively, the electric displacement vector and the magnetic flux density vector, which depend on the applied electric field vector $\mathbf{E}$ (V/m) and on the magnetic field vector $\mathbf{H}$ (A/m) as well as on the material polarizations via the electric polarization density vector $\mathbf{P}$ (C/m^2) and the magnetic polarization density vector $\mathbf{M}$ (A/m). The constants $\epsilon_0 = 8.8541 \times 10^{-12}$ F/m and $\mu_0 = 4\pi \times 10^{-7}$ H/m are the vacuum permittivity and permeability, respectively.

The polarization densities in (2.1) are typically expressed in terms of quantities which are either microscopic, the polarizabilities, or macroscopic, the susceptibilities. Although both the microscopic and macroscopic descriptions are applicable for modeling metasurfaces, we will mostly use the macroscopic model throughout the book, because it more conveniently describes metasurfaces as homogeneous media. For a bianisotropic metasurface, the polarization densities in (2.1) read

$$\mathbf{P} = \epsilon_0 \overline{\overline{\chi}}_{ee} \cdot \mathbf{E} + \frac{1}{c_0} \overline{\overline{\chi}}_{em} \cdot \mathbf{H}, \tag{2.2a}$$

$$\mathbf{M} = \overline{\overline{\chi}}_{mm} \cdot \mathbf{H} + \frac{1}{\eta_0} \overline{\overline{\chi}}_{me} \cdot \mathbf{E}, \tag{2.2b}$$

where $c_0 = 1/\sqrt{\epsilon_0 \mu_0} = 299{,}792{,}458$ m/s and $\eta_0 = \sqrt{\mu_0/\epsilon_0} = 376.73$ Ω are the speed of light in a vacuum and the vacuum impedance, respectively, and $\overline{\overline{\chi}}_{ee}$, $\overline{\overline{\chi}}_{mm}$, $\overline{\overline{\chi}}_{em}$, and $\overline{\overline{\chi}}_{me}$ are the electric, magnetic, magnetic-to-electric, and electric-to-magnetic susceptibility tensors, respectively. Note that these susceptibilities are all unitless. Substituting (2.2) into (2.1) yields the bianisotropic constitutive relations

$$\mathbf{D} = \epsilon_0 \left(\overline{\overline{I}} + \overline{\overline{\chi}}_{ee} \right) \cdot \mathbf{E} + \frac{1}{c_0} \overline{\overline{\chi}}_{em} \cdot \mathbf{H}, \tag{2.3a}$$

$$\mathbf{B} = \mu_0 \left(\overline{\overline{I}} + \overline{\overline{\chi}}_{mm} \right) \cdot \mathbf{H} + \frac{1}{c_0} \overline{\overline{\chi}}_{me} \cdot \mathbf{E}, \tag{2.3b}$$

where $\overline{\overline{I}}$ is the unity dyadic tensor. Sometimes, these relations are also expressed in the more compact form

$$\mathbf{D} = \overline{\overline{\epsilon}} \cdot \mathbf{E} + \overline{\overline{\xi}} \cdot \mathbf{H}, \tag{2.4a}$$

$$\mathbf{B} = \overline{\overline{\zeta}} \cdot \mathbf{E} + \overline{\overline{\mu}} \cdot \mathbf{H}, \tag{2.4b}$$

where $\overline{\overline{\epsilon}}$ (F/m), $\overline{\overline{\mu}}$ (H/m), $\overline{\overline{\xi}}$ (s/m), and $\overline{\overline{\zeta}}$ (s/m) are the permittivity, permeability, magnetic-to-electric, and electric-to-magnetic tensors, respectively.

In basic optical and microwave engineering, the susceptibility tensors in (2.3) typically reduce to scalar quantities because most materials are isotropic. Moreover, the magnetic-to-electric and electric-to-magnetic terms are often ignored because the majority of materials do not induce such kind of magnetoelectric coupling. In contrast, metamaterial technology allows one to engineer artificial materials that exhibit very sophisticated electromagnetic responses involving all the 36 susceptibility components[1] of these tensors.

In addition, the metamaterial susceptibilities may be functions of: (i) the position, $\mathbf{r}$, (ii) the time, t, (iii) the frequency, ω (temporal dispersion), and (iv) the direction of wave propagation, $\mathbf{k}$ (spatial dispersion), or a combination of these dependencies. This suggests a grand classification of metamaterials in terms of direct and Fourier-inverse space and time dependencies, as shown in Figure 2.1, where the 16 distinct types of dependencies correspond to a myriad of distinct bianisotropic media.

Practically, some of the 16 media types represented in Figure 2.1 are still challenging. For instance, time-varying ($\overline{\overline{\chi}}(t)$) or spatially dispersive ($\overline{\overline{\chi}}(\mathbf{k})$) metamaterials are more difficult to realize than spatially varying ones ($\overline{\overline{\chi}}(\mathbf{r})$), which only involve a spatial modulation in their geometry. Moreover, all the materials are de facto temporally dispersive,[2] and particularly metamaterials, which strongly rely on resonant scattering particles to manipulate electromagnetic waves. Finally, some of these dependency combinations are also restricted by the uncertainty principle, as shown in [24]. Thus, metamaterial technology is often limited to a subset of the material types in Figure 2.1. Specifically, the most common types of metamaterials exhibit material parameters such as $\overline{\overline{\chi}}(\omega)$ (e.g. quarter/half-wave plates) and $\overline{\overline{\chi}}(\mathbf{r}, \omega)$ (e.g. lenses and refractors) and, to a lesser extent, $\overline{\overline{\chi}}(\mathbf{k}, \omega)$ (e.g. angularly asymmetric absorbers) and $\overline{\overline{\chi}}(\mathbf{r}, \mathbf{k}, \omega)$ (e.g. diffractionless refractors). It should be emphasized that this rich diversity of space–time variance and dispersion illustrated in Figure 2.1 is restricted to linear metamaterials. As we will see in Section 4.3, the introduction of nonlinearity further increases the number of degrees of freedom for controlling electromagnetic waves.

Throughout the book, we will adopt a particular naming convention to better describe the type of medium we will be dealing with. A medium with $\overline{\overline{\chi}}_{\mathrm{ee}} \neq 0$, $\overline{\overline{\chi}}_{\mathrm{mm}} \neq 0$, and $\overline{\overline{\chi}}_{\mathrm{em}} = \overline{\overline{\chi}}_{\mathrm{me}} = 0$ will be referred to as a homoanisotropic medium, where "homo" indicates same excitation-to-response effects, i.e. electric-

---

1 Each of the four tensors in (2.3) contains $3 \times 3 = 9$ susceptibility components, amounting to a total of $4 \times 9 = 36$ susceptibilities.

2 Note that some materials, such as glass, can often be assumed to be dispersionless within a limited frequency range given the negligible variations of their constitutive parameters across that range.

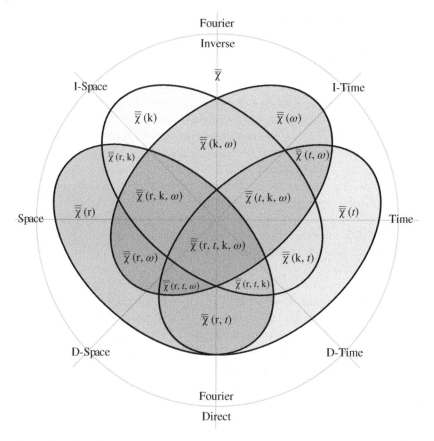

Figure 2.1 Classification of (bianisotropic) metamaterials in terms of their space–time variance and dispersion. The I- and D-notations refer to the inverse and direct space/time domains.

to-electric or magnetic-to-magnetic. A medium with $\overline{\overline{\chi}}_{ee} = \overline{\overline{\chi}}_{mm} = 0$, $\overline{\overline{\chi}}_{em} \neq 0$, and $\overline{\overline{\chi}}_{me} \neq 0$ will be referred to as a heteroanisotropic medium, where "hetero" indicates different excitation-to-response effects, i.e. electric-to-magnetic or magnetic-to-electric. Finally, a medium with nonzero susceptibilities from all four tensors will be generally referred to as a bianisotropic medium.

2.2 Temporal Dispersion

Temporal dispersion is the property according to which the response of a medium, at a given instant of time, depends on the excitation at previous times. This can be mathematically expressed by the time-domain counterparts[3] of (2.3), as

$$\boldsymbol{D}(t) = \epsilon_0 \boldsymbol{\mathcal{E}}(t) + \int_{-\infty}^{T} \left(\epsilon_0 \overline{\overline{\tilde{\chi}}}_{ee}(t-t') \cdot \boldsymbol{\mathcal{E}}(t') + \frac{1}{c_0} \overline{\overline{\tilde{\chi}}}_{em}(t-t') \cdot \boldsymbol{\mathcal{H}}(t') \right) dt',$$
(2.5a)

$$\boldsymbol{B}(t) = \mu_0 \boldsymbol{\mathcal{H}}(t) + \int_{-\infty}^{T} \left(\mu_0 \overline{\overline{\tilde{\chi}}}_{mm}(t-t') \cdot \boldsymbol{\mathcal{H}}(t') + \frac{1}{c_0} \overline{\overline{\tilde{\chi}}}_{me}(t-t') \cdot \boldsymbol{\mathcal{E}}(t') \right) dt',$$
(2.5b)

where only the time dependence of the fields is explicitly mentioned and the spatial dependence is omitted for conciseness. These expressions indicate that the material responses, $\boldsymbol{D}$ and $\boldsymbol{B}$, are the temporal convolution of the susceptibilities and excitation functions. They may be alternatively written as

$$\boldsymbol{D}(t) = \epsilon_0 \boldsymbol{\mathcal{E}}(t) + \epsilon_0 \overline{\overline{\tilde{\chi}}}_{ee}(t) * \boldsymbol{\mathcal{E}}(t) + \frac{1}{c_0} \overline{\overline{\tilde{\chi}}}_{em}(t) * \boldsymbol{\mathcal{H}}(t),$$
(2.6a)

$$\boldsymbol{B}(t) = \mu_0 \boldsymbol{\mathcal{H}}(t) + \mu_0 \overline{\overline{\tilde{\chi}}}_{mm}(t) * \boldsymbol{\mathcal{H}}(t) + \frac{1}{c_0} \overline{\overline{\tilde{\chi}}}_{me}(t) * \boldsymbol{\mathcal{E}}(t).$$
(2.6b)

Since the convolution in the time domain corresponds to the multiplication in the frequency domain, the time-domain Fourier transform of (2.6) corresponds to (2.3).

2.2.1 Causality and Kramers–Kronig Relations

We have mentioned that matter is temporally dispersive and shown that the corresponding medium parameters at a given instant of time depend on the excitation at previous times. Combining these facts with the concept of causality allows one to relate the real and imaginary parts of the material parameters to each other. The corresponding relations are referred to as the Kramers–Kronig relations. Given their crucial importance and given their unclear presentation in the literature, we shall now precisely derive them.

The fundamental tenet of causality is that an effect cannot precede its cause, so that, in particular, matter cannot respond before being excited. Therefore, assuming that a medium starts to be excited at $t = 0$, its response at $t < 0$ must necessarily be zero. We have thus [59]

$$\tilde{\chi}_v(t) = \mathcal{U}(t)\mathcal{A}(t),$$
(2.7)

where $\tilde{\chi}_v(t)$ represents the variation of any of the susceptibility components in (2.3) from the steady-state regime, $\mathcal{U}(t)$ is the Heaviside step function, and $\mathcal{A}(t)$ is an arbitrary function that satisfies $\tilde{\chi}_v(t) = \mathcal{A}(t)$ for $t > 0$. Transforming (2.7) to the

---

3 Here, the tildes are used to differentiate the time-domain susceptibilities from their frequency-domain counterparts.

frequency domain, we get

$$\chi_v(\omega) = \frac{1}{2\pi} U(\omega) * A(\omega), \tag{2.8}$$

where

$$U(\omega) = \mathcal{F}\{\mathcal{U}(t)\} = \text{PV}\left(\frac{j}{\omega}\right) + \pi\delta(\omega), \tag{2.9}$$

with $\mathcal{F}\{\cdot\}$ representing the Fourier transformation operation, $\delta(\omega)$ being the Dirac delta distribution, and $\text{PV}(\cdot)$ denoting the principal value of its argument. Substituting (2.9) into (2.8) leads to the explicit convolution relation

$$
\begin{aligned}
\chi_v(\omega) &= \frac{1}{2\pi} \int_{-\infty}^{+\infty} \left(\text{PV}\left(\frac{j}{\omega - \omega'}\right) + \pi\delta(\omega - \omega') \right) A(\omega') d\omega' \\
&= \frac{1}{2\pi} \text{PV} \int_{-\infty}^{+\infty} \frac{jA(\omega')}{\omega - \omega'} d\omega' + \frac{1}{2}A(\omega).
\end{aligned}
\tag{2.10}
$$

Noting that $\chi_v(\omega)$ is a complex quantity, we can rewrite (2.10) in the more explicit form

$$\chi_{v,r}(\omega) + j\chi_{v,i}(\omega) = \frac{1}{2\pi} \text{PV} \int_{-\infty}^{+\infty} \frac{jA(\omega')}{\omega - \omega'} d\omega' + \frac{1}{2}A(\omega), \tag{2.11}$$

where $\chi_{v,r}(\omega)$ and $\chi_{v,i}(\omega)$ are the real and imaginary parts of χ_v, respectively. Since $A(t)$ may take arbitrary values for $t < 0$, we chose $A(t)$ such that it is time-symmetric (or even), i.e. $A(t) = A(-t)$, which leads to a purely real function $A(\omega)$. We can then separate (2.11) into its real and imaginary parts as

$$\chi_{v,r}(\omega) = \frac{1}{2}A(\omega), \tag{2.12a}$$

$$\chi_{v,i}(\omega) = \frac{1}{2\pi} \text{PV} \int_{-\infty}^{+\infty} \frac{A(\omega')}{\omega - \omega'} d\omega'. \tag{2.12b}$$

Substituting (2.12a) into (2.12b) yields

$$\chi_{v,i}(\omega) = \frac{1}{\pi} \text{PV} \int_{-\infty}^{+\infty} \frac{\chi_{v,r}(\omega')}{\omega - \omega'} d\omega'. \tag{2.13}$$

Following a similar procedure with a time-asymmetric $A(t)$ function, i.e. $A(t) = -A(-t)$, corresponding to a purely imaginary $A(\omega)$, we obtain

$$\chi_{v,r}(\omega) = -\frac{1}{\pi} \text{PV} \int_{-\infty}^{+\infty} \frac{\chi_{v,i}(\omega')}{\omega - \omega'} d\omega'. \tag{2.14}$$

Equations (2.13) and (2.14) are the sought Kramers–Kronig relations. They indicate that variations in the real part of the medium parameters are *necessarily* associated with specific variations in their imaginary parts. For a quantity such as the refractive index, for instance, this implies that transforming an electromagnetic wave (e.g. negatively refracting it) from the real part of the index (e.g. making it negative) necessarily results in a variation of the imaginary part of the index, which corresponds to inserting loss, and vice versa.

2.2.2 Lorentz Oscillator Model

The Lorentz oscillator is a simple classical atomic model that describes the interaction of a time-harmonic field with matter. It was initially developed to describe the resonant behavior of an electron cloud. This model has been widely used to describe the temporal-frequency dispersive nature of materials and their related frequency-dependent refractive index [140]. It turns out to be also particularly useful in describing the responses of metamaterials, as they are generally made of resonant scattering particles.

Let us consider that an electron cloud is subjected to the Lorentz electric force,

$$\mathbf{\mathcal{F}}_e(\mathbf{r}, t) = -q_e \mathbf{\mathcal{E}}_{loc}(\mathbf{r}, t), \tag{2.15}$$

where q_e is the electric charge and $\mathbf{\mathcal{E}}_{loc}(\mathbf{r}, t)$ is the local field.[4] We assume here that the magnetic force is negligible compared to the electric force, which is the case for nonrelativistic velocities, and that the nuclei, which are much heavier than the electrons, are not moving. The restoring force between the nuclei and the electrons can be expressed similarly to the force of a mass attached to a spring, i.e.

$$\mathbf{\mathcal{F}}_r(\mathbf{r}, t) = -m_e \omega_r^2 \mathbf{d}(\mathbf{r}, t), \tag{2.16}$$

where m_e is the mass of the electron cloud, ω_r is a constant analogous to the stiffness of the spring, and $\mathbf{d}(\mathbf{r}, t)$ is the displacement from equilibrium of the electron cloud. Finally, to model dissipation, we introduce the frictional force

$$\mathbf{\mathcal{F}}_f(\mathbf{r}, t) = -2\Gamma m_e \mathbf{v}(\mathbf{r}, t), \tag{2.17}$$

where $\mathbf{v}(\mathbf{r}, t)$ is the displacement velocity of the electron cloud and Γ is a constant representing the friction coefficient. Applying Newton's second law with these three forces, we obtain

$$m_e \frac{\partial}{\partial t} \mathbf{v}(\mathbf{r}, t) = -q_e \mathbf{\mathcal{E}}_{loc}(\mathbf{r}, t) - m_e \omega_r^2 \mathbf{d}(\mathbf{r}, t) - 2\Gamma m_e \mathbf{v}(\mathbf{r}, t). \tag{2.18}$$

Rearranging the terms and noting that $\mathbf{v}(\mathbf{r}, t) = \partial \mathbf{d}(\mathbf{r}, t)/\partial t$, we get

$$\frac{\partial^2}{\partial t^2} \mathbf{d}(\mathbf{r}, t) + 2\Gamma \frac{\partial}{\partial t} \mathbf{d}(\mathbf{r}, t) + \omega_r^2 \mathbf{d}(\mathbf{r}, t) = -\frac{q_e}{m_e} \mathbf{\mathcal{E}}_{loc}(\mathbf{r}, t). \tag{2.19}$$

This equation may be written in a more convenient form by noting that the displacement from equilibrium is related to the electric polarization density as

---

4 The local field is defined as the total field, at the position of the particle, minus the field produced by the particle itself.

$P(\mathbf{r}, t) = -q_e N \mathbf{d}(\mathbf{r}, t)$ [140], where N is the electron density. This transforms (2.19) into

$$\frac{\partial^2}{\partial t^2} P(\mathbf{r}, t) + 2\Gamma \frac{\partial}{\partial t} P(\mathbf{r}, t) + \omega_r^2 P(\mathbf{r}, t) = \frac{N q_e^2}{m_e} \mathcal{E}_{\text{loc}}(\mathbf{r}, t). \tag{2.20}$$

For electrically small particles, we may use the Clausius–Mosotti expression, which relates the local field to the total electric, $\mathcal{E}(\mathbf{r}, t)$ and to the polarization density as $\mathcal{E}_{\text{loc}}(\mathbf{r}, t) = \mathcal{E}(\mathbf{r}, t) + P(\mathbf{r}, t)/(3\epsilon_0)$ [140], which reduces (2.20) into

$$\frac{\partial^2}{\partial t^2} P(\mathbf{r}, t) + 2\Gamma \frac{\partial}{\partial t} P(\mathbf{r}, t) + \omega_0^2 P(\mathbf{r}, t) = \frac{N q_e^2}{m_e} \mathcal{E}(\mathbf{r}, t), \tag{2.21}$$

where $\omega_0 = \sqrt{\omega_r^2 - N q_e^2/(3 m_e \epsilon_0)}$ is the resonant frequency. In the harmonic regime, this equation becomes

$$-\omega^2 \mathbf{P} + j2\omega\Gamma \mathbf{P} + \omega_0^2 \mathbf{P} = \frac{N q_e^2}{m_e} \mathbf{E}, \tag{2.22}$$

where $\mathbf{P}$ is the time-domain Fourier transform of $P(\mathbf{r}, t)$ and $\mathbf{E}$ that of $\mathcal{E}(\mathbf{r}, t)$. Finally substituting $\mathbf{P} = \epsilon_0 \chi_{ee} \mathbf{E}$, eliminating $\mathbf{E}$, and solving for χ_{ee} yields the dispersive susceptibility

$$\chi_{ee}(\omega) = \frac{\omega_p^2}{\omega_0^2 - \omega^2 + j2\omega\Gamma}, \tag{2.23}$$

where $\omega_p = \sqrt{N q_e^2/(\epsilon_0 m_e)}$ is the plasma frequency. Equation (2.23) is referred to as the Lorentz model and its frequency dependence is plotted in Figure 2.2. Note that the susceptibility (2.23) is a complex quantity that naturally satisfies Kramers–Kronig relations (Eqs. (2.13) and (2.14)).

An important particular case of the Lorentz model (2.23) is the Drude model, which applies to metals. In a metal, the electrons are free charge carriers, which implies that the restoring force (2.16) is zero, leading to $\omega_0 = 0$ in (2.23). The Drude model is particularly useful for modeling metals in the optical regime, where the frequency of light approaches the plasma frequency.

The plasma frequency and damping of aluminum, gold, and silver, which are metals frequently used in the optical regime, are reported in Table 2.1. As the frequency increases and approaches the plasma frequency, these metals behave more and more as lossy dielectrics. In contrast, in the microwave regime, metals behave as perfect electric conductors (PEC), with negligible dispersion. This is of particular importance for the practical realization of metallic-based metasurfaces, especially in the optical regime where the lossy nature of these metals becomes substantial.

To further illustrate the dispersive behavior of these metals, Figure 2.3 plots the real and imaginary parts of their permittivity in the optical regime. At the longer

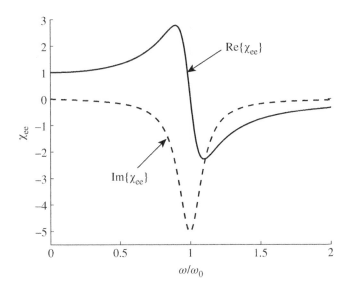

Figure 2.2 Dispersive response of the electric susceptibility of a resonant structure for the parameter $\omega_p/\Gamma = 10$ (Eq. (2.23)).

Table 2.1 Plasma frequency ω_p (corresponding wavelength) and damping Γ for three common metals [116].

Metal	Plasma frequency, ω_p (eV)	Damping, Γ (eV)
Ag	9.013 (137.56 nm)	0.018
Au	9.026 (137.36 nm)	0.0267
Al	14.75 (84.06 nm)	0.0818

wavelengths, they behave as expected, i.e. they exhibit a permittivity with large negative real and imaginary parts. However, as the wavelength tends toward the plasma frequency, their optical behavior changes: they become less opaque and thus let light penetrate deeper in them.

2.3 Spatial Dispersion

In addition of being temporally dispersive, a medium may also be spatially dispersive. As temporal dispersion is a temporally nonlocal phenomenon, spatial

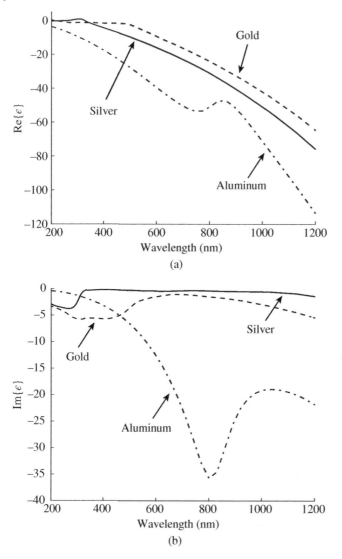

Figure 2.3 Experimental dispersion curves of the permittivity of silver, gold, and aluminum [72, 102]. (a) Real part. (b) Imaginary part.

dispersion is a spatially nonlocal phenomenon, whereby the response of the medium at the position **r** depends on its excitation at position **r′** [9, 29, 52, 148]. This effect is important in metamaterials because it is fundamentally related to bianisotropy and artificial magnetism, as we shall now demonstrate.

Note that an extensive treatment of the topic of spatial dispersion would be beyond the scope of this book. Here, we thus limit ourselves to a brief and simplified description of this phenomenon, while more advanced presentations may be found in [9, 29, 52, 148].

In order to show how spatial dispersion brings about bianisotropy and artificial magnetism in the constitutive relations, consider a medium with the conventional constitutive relations

$$\mathbf{D} = \epsilon_0 \mathbf{E} + \mathbf{P} = \epsilon_0 \mathbf{E} + \frac{1}{j\omega} \mathbf{J}_{\text{ind}}, \tag{2.24a}$$

$$\mathbf{H} = \mu_0^{-1} \mathbf{B}, \tag{2.24b}$$

where we have expressed the material polarization density, $\mathbf{P}$, in terms of the induced current density, $\mathbf{J}_{\text{ind}}$ (A/m^2). Spatial dispersion is now introduced via the relationship between the induced current and the electric field [29],

$$\mathbf{J}_{\text{ind}}(\mathbf{r}) = \int_V \overline{\overline{K}}(\mathbf{r}, \mathbf{r}') \cdot \mathbf{E}(\mathbf{r}') dV, \tag{2.25}$$

where the dyadic tensor $\overline{\overline{K}}$ represents the response of the medium and V is a volume of integration, centered at $\mathbf{r}$. We next restrict our attention to the case of weak spatial dispersion, where $V \ll \lambda$. We can then simplify (2.25) by using the three-dimensional Taylor expansion of $\mathbf{E}(\mathbf{r}')$ truncated to the second order [29],

$$\mathbf{E}(\mathbf{r}') = \mathbf{E}(\mathbf{r}) + \left(\nabla_\alpha \mathbf{E}\right)|_{\mathbf{r}}(r'_\alpha - r_\alpha) + \frac{1}{2}\left(\nabla_\beta \nabla_\alpha \mathbf{E}\right)|_{\mathbf{r}}(r'_\alpha - r_\alpha)(r'_\beta - r_\beta), \tag{2.26}$$

where the subscripts α and β run over x, y, and z, which decomposes the second term in three terms and the third term in nine terms. Substituting (2.26) into (2.25) leads, after somewhat involved manipulations and simplifications [29], to

$$J_{\text{ind},i} = j\omega \left[a_{ij} E_j + a_{ijk} \left(\nabla_k E_j\right) + a_{ijkl} \left(\nabla_l \nabla_k E_j\right) \right], \tag{2.27}$$

where the parameters $a_{...}$ are related to $\overline{\overline{K}}$. Inserting then (2.27) into 2.24a yields

$$D_i = \left(\epsilon_0 \delta_{ij} + a_{ij}\right) E_j + a_{ijk} \left(\nabla_k E_j\right) + a_{ijkl} \left(\nabla_l \nabla_k E_j\right) \tag{2.28}$$

where δ_{ij} is the Kronecker delta. The relation (2.28) corresponds to general second-order weak spatial dispersion constitutive relations. The presence of the spatial derivatives makes it practically cumbersome, and we shall therefore transform it into a simpler form [148]. For simplicity, we restrict our attention to the isotropic version of (2.28), which may be written as [29]

$$\mathbf{D} = \epsilon \mathbf{E} + \alpha \nabla \times \mathbf{E} + \beta \nabla \nabla \cdot \mathbf{E} + \gamma \nabla \times \nabla \times \mathbf{E}, \tag{2.29}$$

where ϵ is the permittivity, and α, β, and γ are complex scalar constants related to the parameters $a_{...}$ in (2.28). In order to simplify (2.29), we shall first demonstrate that Maxwell equations are invariant under the transformation [148]

$$\mathbf{D}' = \mathbf{D} + \nabla \times \mathbf{Q}, \tag{2.30a}$$

$$\mathbf{H}' = \mathbf{H} + j\omega \mathbf{Q}, \tag{2.30b}$$

where $\mathbf{Q}$ is an arbitrary vector. To prove this invariance, consider Maxwell–Ampère equation

$$\nabla \times \mathbf{H} = j\omega\mathbf{D}. \tag{2.31}$$

Substituting (2.30) into this relation yields

$$\nabla \times \mathbf{H}' - j\omega\nabla \times \mathbf{Q} = j\omega\mathbf{D}' - j\omega\nabla \times \mathbf{Q}, \tag{2.32}$$

which clearly reduces to

$$\nabla \times \mathbf{H}' = j\omega\mathbf{D}', \tag{2.33}$$

hence proving the equivalence of (2.33) and (2.31), and therefore demonstrating the invariance of Maxwell equations under the transformation (2.30) for any $\mathbf{Q}$ [148]. We now substitute (2.29) into (2.30) along with $\mathbf{Q}_1 = -\frac{\alpha}{2}\mathbf{E}$, which yields

$$\mathbf{D}' = \mathbf{D} + \nabla \times \mathbf{Q}_1 = \epsilon\mathbf{E} + \frac{\alpha}{2}\nabla \times \mathbf{E} + \beta\nabla\nabla \cdot \mathbf{E} + \gamma\nabla \times \nabla \times \mathbf{E}, \tag{2.34a}$$

$$\mathbf{H}' = \mathbf{H} + j\omega\mathbf{Q}_1 = \mu_0^{-1}\mathbf{B} - j\omega\frac{\alpha}{2}\mathbf{E}. \tag{2.34b}$$

Further substituting Maxwell–Faraday equation, $\nabla \times \mathbf{E} = -j\omega\mathbf{B}$, into (2.34) yields

$$\mathbf{D}' = \epsilon\mathbf{E} - j\omega\frac{\alpha}{2}\mathbf{B} + \beta\nabla\nabla \cdot \mathbf{E} + \nabla \times (-j\omega\gamma\mathbf{B}), \tag{2.35a}$$

$$\mathbf{H}' = \mu_0^{-1}\mathbf{B} - j\omega\frac{\alpha}{2}\mathbf{E}. \tag{2.35b}$$

Finally substituting these relations with $\mathbf{Q}_2 = -j\omega\gamma\mathbf{B}$ into (2.30) leads to the relations

$$\mathbf{D} = \mathbf{D}' - \nabla \times \mathbf{Q}_2 = \epsilon\mathbf{E} - j\omega\frac{\alpha}{2}\mathbf{B} + \beta\nabla\nabla \cdot \mathbf{E}, \tag{2.36a}$$

$$\mathbf{H} = \mathbf{H} - j\omega\mathbf{Q}_2 = \mu_0^{-1}\mathbf{B} - j\omega\frac{\alpha}{2}\mathbf{E} - \omega^2\gamma\mathbf{B}, \tag{2.36b}$$

which take the compact form

$$\mathbf{D} = \epsilon\mathbf{E} - jv\mathbf{B} + \beta\nabla\nabla \cdot \mathbf{E}, \tag{2.37a}$$

$$\mathbf{H} = \mu^{-1}\mathbf{B} - jv\mathbf{E}, \tag{2.37b}$$

where $v = \omega\frac{\alpha}{2}$ is a coupling parameter associated to chirality [26, 155] and where the parameter $\mu = \mu_0/(1 - \omega^2\mu_0\gamma)$ corresponds to artificial magnetism [122]. The constitutive relations (2.37) reveal, via (2.29), that chirality is related to spatial dispersion of the first order via α in v, while artificial magnetism is related to spatial dispersion of the second order via γ. Both chirality and artificial magnetism depend on the excitation frequency via ω. It may seem surprising that artificial magnetism is related to spatial dispersion but consider the following simple example, which is one of the easiest ways of creating an effective magnetic dipole.

Consider, two small metal strips, one placed at a subwavelength distance d to the other in the direction of wave propagation. If the conditions are met, a mode with an odd current distribution may be excited on the strips resulting in an effective magnetic dipole. In that case, the electric field on one of the strip slightly differs from the one on the other strip since they are separated by a distance d, thus implying that the effective magnetic response of the strips spatially depends on the exciting electric field. If $d \rightarrow 0$, then the difference of the electric field on the strips would vanish as would the effective magnetic dipole moment.

It is not possible to further transform (2.37a) to eliminate the double spatial derivative associated with β. However, it turns out that β is most often negligible compared to v and γ, as discussed in [29, 148], so that this term may generally be ignored. Solving then (2.37) for **B** ultimately leads to the spatial-dispersion relations

$$\mathbf{D} = \epsilon' \mathbf{E} + \xi \mathbf{H}, \tag{2.38a}$$

$$\mathbf{B} = \zeta \mathbf{E} + \mu \mathbf{H}, \tag{2.38b}$$

where $\zeta = jv\mu$, $\xi = -jv\mu$, and $\epsilon' = \epsilon + \mu v^2$, which are the isotropic (or biisotropic) counterparts of the bianisotropic constitutive relations (2.4).

Note that when $\omega \rightarrow 0$, then $\mu \rightarrow \mu_0$ and $\epsilon' \rightarrow \epsilon \neq \epsilon_0$. This means that the permittivity and permeability have fundamentally different dispersive nature, a fact that has important consequences for metamaterials, as we will see throughout the book.

2.4 Lorentz Reciprocity Theorem

Reciprocity or nonreciprocity is a fundamental physical property of all media, structures, devices, and systems. "A nonreciprocal (reciprocal) system is defined as a system that exhibits different (same) received–transmitted field ratios when its source(s) and detector(s) are exchanged" [28].

A linear-time-invariant (LTI) system can be made nonreciprocal *only* via the application of an external bias field that is odd under time reversal, such as a magnetic field or a current. The most common example of a nonreciprocal device is the Faraday isolator, whose nonreciprocity is obtained by biasing a ferrite with an external static magnetic field [143].

This section derives the Lorentz reciprocity theorem for a bianisotropic LTI medium, which provides the general conditions for reciprocity in terms of susceptibility tensors. These relations naturally apply to LTI metasurfaces as well.

Let us consider a volume V enclosed by a surface S and containing the impressed electric and magnetic sources $\mathbf{J}$ and $\mathbf{K}$, respectively. In the time-harmonic regime, these sources are related to the fields $\mathbf{E}$, $\mathbf{D}$, $\mathbf{H}$, and $\mathbf{B}$ via Maxwell equations,

$$\nabla \times \mathbf{E} = -\mathbf{K} - j\omega\mathbf{B}, \tag{2.39a}$$

$$\nabla \times \mathbf{H} = \mathbf{J} + j\omega\mathbf{D}. \tag{2.39b}$$

Similarly, the phase-conjugated impressed sources,[5] $\mathbf{J}^*$ and $\mathbf{K}^*$, are related to the phase-conjugated fields $\mathbf{E}^*$, $\mathbf{D}^*$, $\mathbf{H}^*$, and $\mathbf{B}^*$ as

$$\nabla \times \mathbf{E}^* = -\mathbf{K}^* - j\omega\mathbf{B}^*, \tag{2.40a}$$

$$\nabla \times \mathbf{H}^* = \mathbf{J}^* + j\omega\mathbf{D}^*. \tag{2.40b}$$

Subtracting 2.39a pre-multiplied by $\mathbf{H}^*$ from 2.40b pre-multiplied by $\mathbf{E}$, and subtracting 2.40a pre-multiplied by $\mathbf{H}$ from 2.39b pre-multiplied by $\mathbf{E}^*$ yields

$$\mathbf{E} \cdot \nabla \times \mathbf{H}^* - \mathbf{H}^* \cdot \nabla \times \mathbf{E} = \mathbf{E} \cdot \mathbf{J}^* + j\omega\mathbf{E} \cdot \mathbf{D}^* + \mathbf{H}^* \cdot \mathbf{K} + j\omega\mathbf{H}^* \cdot \mathbf{B}, \tag{2.41a}$$

$$\mathbf{E}^* \cdot \nabla \times \mathbf{H} - \mathbf{H} \cdot \nabla \times \mathbf{E}^* = \mathbf{E}^* \cdot \mathbf{J} + j\omega\mathbf{E}^* \cdot \mathbf{D} + \mathbf{H} \cdot \mathbf{K}^* + j\omega\mathbf{H} \cdot \mathbf{B}^*. \tag{2.41b}$$

Using the identity $\mathbf{A} \cdot \nabla \times \mathbf{B} - \mathbf{B} \cdot \nabla \times \mathbf{A} = -\nabla \cdot (\mathbf{A} \times \mathbf{B})$ reduces these equations to

$$-\nabla \cdot (\mathbf{E} \times \mathbf{H}^*) = \mathbf{E} \cdot \mathbf{J}^* + j\omega\mathbf{E} \cdot \mathbf{D}^* + \mathbf{H}^* \cdot \mathbf{K} + j\omega\mathbf{H}^* \cdot \mathbf{B}, \tag{2.42a}$$

$$-\nabla \cdot (\mathbf{E}^* \times \mathbf{H}) = \mathbf{E}^* \cdot \mathbf{J} + j\omega\mathbf{E}^* \cdot \mathbf{D} + \mathbf{H} \cdot \mathbf{K}^* + j\omega\mathbf{H} \cdot \mathbf{B}^*. \tag{2.42b}$$

Subtracting 2.42a from 2.42b yields

$$\nabla \cdot (\mathbf{E} \times \mathbf{H}^* - \mathbf{E}^* \times \mathbf{H}) = j\omega \, (\mathbf{E}^* \cdot \mathbf{D} - \mathbf{E} \cdot \mathbf{D}^* + \mathbf{H} \cdot \mathbf{B}^* - \mathbf{H}^* \cdot \mathbf{B})$$
$$+ \mathbf{E}^* \cdot \mathbf{J} - \mathbf{E} \cdot \mathbf{J}^* + \mathbf{H} \cdot \mathbf{K}^* - \mathbf{H}^* \cdot \mathbf{K}. \tag{2.43}$$

We may now integrate this equation over the volume V and apply the Gauss theorem, which gives

$$\oiint_S (\mathbf{E} \times \mathbf{H}^* - \mathbf{E}^* \times \mathbf{H}) \cdot d\mathbf{S}$$
$$= j\omega \iiint_V (\mathbf{E}^* \cdot \mathbf{D} - \mathbf{E} \cdot \mathbf{D}^* + \mathbf{H} \cdot \mathbf{B}^* - \mathbf{H}^* \cdot \mathbf{B}) \, dV$$
$$+ \iiint_V (\mathbf{E}^* \cdot \mathbf{J} - \mathbf{E} \cdot \mathbf{J}^* + \mathbf{H} \cdot \mathbf{K}^* - \mathbf{H}^* \cdot \mathbf{K}) \, dV. \tag{2.44}$$

---

5 In the time domain, time reversal consists in replacing the variable t by $-t$. Therefore, a function $f(t)$ becomes $f(-t)$. In the frequency domain, the time-reversal operation consists in taking the complex conjugate of the Fourier transform of $f(t)$, i.e. in transforming $\tilde{f}(\omega)$ into $\pm\tilde{f}(\omega)^*$, where the plus sign is for a time-even quantity while the minus sign is for a time-odd quantity [28].

The surface integral on the left-hand side of (2.44) and the second volume integral on its right-hand side both vanish when integrated over an unbounded medium [28, 81]. This results into

$$\iiint_V (\mathbf{E}^* \cdot \mathbf{D} - \mathbf{E} \cdot \mathbf{D}^* + \mathbf{H} \cdot \mathbf{B}^* - \mathbf{H}^* \cdot \mathbf{B}) \, dV = 0, \tag{2.45}$$

which represents a fundamental relation for reciprocity. We next include the bianisotropic material parameters in this relation. For this purpose, we use the bianisotropic constitutive relations (2.4), which may be explicitly rewritten as

$$\mathbf{D} = \bar{\bar{\epsilon}}(\mathbf{F}_0) \cdot \mathbf{E} + \bar{\bar{\xi}}(\mathbf{F}_0) \cdot \mathbf{H}, \tag{2.46a}$$

$$\mathbf{B} = \bar{\bar{\zeta}}(\mathbf{F}_0) \cdot \mathbf{E} + \bar{\bar{\mu}}(\mathbf{F}_0) \cdot \mathbf{H}, \tag{2.46b}$$

where $\mathbf{F}_0$ represents a time-odd external bias. Under phase conjugation (or time reversal), these constitutive relations become [28]

$$\mathbf{D}^* = \bar{\bar{\epsilon}}(-\mathbf{F}_0) \cdot \mathbf{E}^* + \bar{\bar{\xi}}(-\mathbf{F}_0) \cdot \mathbf{H}^*, \tag{2.47a}$$

$$\mathbf{B}^* = \bar{\bar{\zeta}}(-\mathbf{F}_0) \cdot \mathbf{E}^* + \bar{\bar{\mu}}(-\mathbf{F}_0) \cdot \mathbf{H}^*. \tag{2.47b}$$

Substituting (2.46) and (2.47) into (2.45) transforms the integrand of (2.45) into

$$\mathbf{E}^* \cdot \left[\bar{\bar{\epsilon}}(\mathbf{F}_0) \cdot \mathbf{E} + \bar{\bar{\xi}}(\mathbf{F}_0) \cdot \mathbf{H} \right] - \mathbf{E} \cdot \left[\bar{\bar{\epsilon}}(-\mathbf{F}_0) \cdot \mathbf{E}^* + \bar{\bar{\xi}}(-\mathbf{F}_0) \cdot \mathbf{H}^* \right]$$
$$+ \mathbf{H} \cdot \left[\bar{\bar{\zeta}}(-\mathbf{F}_0) \cdot \mathbf{E}^* + \bar{\bar{\mu}}(-\mathbf{F}_0) \cdot \mathbf{H}^* \right] - \mathbf{H}^* \cdot \left[\bar{\bar{\zeta}}(\mathbf{F}_0) \cdot \mathbf{E} + \bar{\bar{\mu}}(\mathbf{F}_0) \cdot \mathbf{H} \right]. \tag{2.48}$$

Expanding the bracketed terms in (2.48), using the identity $\mathbf{A} \cdot \bar{\bar{\chi}} \cdot \mathbf{B} = (\mathbf{A} \cdot \bar{\bar{\chi}} \cdot \mathbf{B})^{\mathrm{T}} = \mathbf{B} \cdot \bar{\bar{\chi}}^{\mathrm{T}} \cdot \mathbf{A}$, where the superscript T denotes the transpose operation, and grouping the identical parameter terms, we obtain

$$\mathbf{E}^* \cdot \left[\bar{\bar{\epsilon}}(\mathbf{F}_0) - \bar{\bar{\epsilon}}^{\mathrm{T}}(-\mathbf{F}_0) \right] \cdot \mathbf{E} - \mathbf{H}^* \cdot \left[\bar{\bar{\mu}}(\mathbf{F}_0) - \bar{\bar{\mu}}^{\mathrm{T}}(-\mathbf{F}_0) \right] \cdot \mathbf{H}$$
$$+ \mathbf{E}^* \cdot \left[\bar{\bar{\xi}}(\mathbf{F}_0) + \bar{\bar{\zeta}}^{\mathrm{T}}(-\mathbf{F}_0) \right] \cdot \mathbf{H} - \mathbf{H}^* \cdot \left[\bar{\bar{\zeta}}(\mathbf{F}_0) + \bar{\bar{\xi}}^{\mathrm{T}}(-\mathbf{F}_0) \right] \cdot \mathbf{E}. \tag{2.49}$$

The nullity of the volume integral of (2.49) (Eq. (2.45)) for any field implies then the general[6] bianisotropic Lorentz reciprocity conditions

$$\bar{\bar{\epsilon}}(\mathbf{F}_0) = \bar{\bar{\epsilon}}^{\mathrm{T}}(-\mathbf{F}_0), \quad \bar{\bar{\mu}}(\mathbf{F}_0) = \bar{\bar{\mu}}^{\mathrm{T}}(-\mathbf{F}_0), \tag{2.50a}$$

$$\bar{\bar{\xi}}(\mathbf{F}_0) = -\bar{\bar{\zeta}}^{\mathrm{T}}(-\mathbf{F}_0), \quad \bar{\bar{\zeta}}(\mathbf{F}_0) = -\bar{\bar{\xi}}^{\mathrm{T}}(-\mathbf{F}_0). \tag{2.50b}$$

In the absence of external bias, i.e. $\mathbf{F}_0 = 0$, these conditions reduce to

$$\bar{\bar{\epsilon}} = \bar{\bar{\epsilon}}^{\mathrm{T}}, \quad \bar{\bar{\mu}} = \bar{\bar{\mu}}^{\mathrm{T}}, \quad \bar{\bar{\xi}} = -\bar{\bar{\zeta}}^{\mathrm{T}}, \tag{2.51}$$

---

6 These relations not only provide the usual reciprocity conditions, but also the *nonreciprocity* conditions [28], e.g. $\bar{\bar{\mu}}(\mathbf{F}_0) \neq \bar{\bar{\mu}}^{\mathrm{T}}(+\mathbf{F}_0)$ in the case of a magnetic nonreciprocal medium.

which may be expressed in terms of susceptibilities as

$$\overline{\overline{\chi}}_{ee} = \overline{\overline{\chi}}_{ee}^T, \quad \overline{\overline{\chi}}_{mm} = \overline{\overline{\chi}}_{mm}^T, \quad \overline{\overline{\chi}}_{me} = -\overline{\overline{\chi}}_{em}^T. \tag{2.52}$$

2.5 Poynting Theorem

This section derives the time-average bianisotropic Poynting theorem [16, 81, 140], which provides the general conditions for gain and loss in terms of susceptibility tensors.

The time-domain Maxwell–Faraday and Maxwell–Ampère equations, assuming the presence of electric current sources, $\boldsymbol{J}$, and magnetic current sources, $\boldsymbol{\mathcal{K}}$, are, respectively, given as

$$\nabla \times \boldsymbol{\mathcal{E}} = -\mathbf{K} - \frac{\partial \boldsymbol{B}}{\partial t}, \tag{2.53a}$$

$$\nabla \times \boldsymbol{\mathcal{H}} = \boldsymbol{J} + \frac{\partial \boldsymbol{D}}{\partial t}, \tag{2.53b}$$

with the bianisotropic constitutive relations (2.4) defined by

$$\boldsymbol{B} = \mu_0(\boldsymbol{\mathcal{H}} + \boldsymbol{\mathcal{M}}), \quad \boldsymbol{\mathcal{M}} = \overline{\overline{\chi}}_{mm} \cdot \boldsymbol{\mathcal{H}} + \frac{1}{\eta_0}\overline{\overline{\chi}}_{me} \cdot \boldsymbol{\mathcal{E}}, \tag{2.54a}$$

$$\boldsymbol{D} = \epsilon_0 \boldsymbol{\mathcal{E}} + \boldsymbol{P}, \quad \boldsymbol{P} = \epsilon_0 \overline{\overline{\chi}}_{ee} \cdot \boldsymbol{\mathcal{E}} + \frac{1}{c_0}\overline{\overline{\chi}}_{em} \cdot \boldsymbol{\mathcal{H}}. \tag{2.54b}$$

Pre-multiplying (2.53a) by $\boldsymbol{\mathcal{H}}$ and (2.53b) by $\boldsymbol{\mathcal{E}}$ and subtracting the two resulting equations yields

$$\boldsymbol{\mathcal{H}} \cdot (\nabla \times \boldsymbol{\mathcal{E}}) - \boldsymbol{\mathcal{E}} \cdot (\nabla \times \boldsymbol{\mathcal{H}}) = -\boldsymbol{\mathcal{E}} \cdot \boldsymbol{J} - \boldsymbol{\mathcal{H}} \cdot \boldsymbol{\mathcal{K}} - \boldsymbol{\mathcal{E}} \cdot \frac{\partial \boldsymbol{D}}{\partial t} - \boldsymbol{\mathcal{H}} \cdot \frac{\partial \boldsymbol{B}}{\partial t}. \tag{2.55}$$

The left-hand side of (2.55) may be simplified using the vectorial identity

$$\boldsymbol{\mathcal{H}} \cdot (\nabla \times \boldsymbol{\mathcal{E}}) - \boldsymbol{\mathcal{E}} \cdot (\nabla \times \boldsymbol{\mathcal{H}}) = \nabla \cdot (\boldsymbol{\mathcal{E}} \times \boldsymbol{\mathcal{H}}), \tag{2.56}$$

where the cross product $\boldsymbol{\mathcal{E}} \times \boldsymbol{\mathcal{H}}$ corresponds to the Poynting vector $\boldsymbol{\mathcal{S}}$. This transforms (2.55) into

$$\nabla \cdot \boldsymbol{\mathcal{S}} = -\boldsymbol{\mathcal{E}} \cdot \boldsymbol{J} - \boldsymbol{\mathcal{H}} \cdot \boldsymbol{\mathcal{K}} - \boldsymbol{\mathcal{E}} \cdot \frac{\partial \boldsymbol{D}}{\partial t} - \boldsymbol{\mathcal{H}} \cdot \frac{\partial \boldsymbol{B}}{\partial t}. \tag{2.57}$$

We shall now simplify the last two terms of this relation to provide the final form of the bianisotropic Poynting theorem. We show the derivations only for $-\boldsymbol{\mathcal{E}} \cdot \frac{\partial \boldsymbol{D}}{\partial t}$, but similar developments can be made for $-\boldsymbol{\mathcal{H}} \cdot \frac{\partial \boldsymbol{B}}{\partial t}$. From $\boldsymbol{D} = \epsilon_0 \boldsymbol{\mathcal{E}} + \boldsymbol{P}$, we have that

$$-\boldsymbol{\mathcal{E}} \cdot \frac{\partial \boldsymbol{D}}{\partial t} = -\boldsymbol{\mathcal{E}} \cdot \frac{\partial}{\partial t}(\epsilon_0 \boldsymbol{\mathcal{E}} + \boldsymbol{P}) = -\epsilon_0 \boldsymbol{\mathcal{E}} \cdot \frac{\partial}{\partial t}\boldsymbol{\mathcal{E}} - \boldsymbol{\mathcal{E}} \cdot \frac{\partial}{\partial t}\boldsymbol{P}. \tag{2.58}$$

Splitting the two terms of the right-hand side into two equal parts transforms this relation into

$$-\mathcal{E} \cdot \frac{\partial D}{\partial t} = -\frac{1}{2}\epsilon_0 \mathcal{E} \cdot \frac{\partial}{\partial t}\mathcal{E} - \frac{1}{2}\mathcal{E} \cdot \frac{\partial}{\partial t}P - \frac{1}{2}\epsilon_0 \mathcal{E} \cdot \frac{\partial}{\partial t}\mathcal{E} - \frac{1}{2}\mathcal{E} \cdot \frac{\partial}{\partial t}P. \tag{2.59}$$

Manipulating the terms in the right-hand side of this new relation, adding the extra null term $\frac{1}{2}P \cdot \frac{\partial}{\partial t}\mathcal{E} - \frac{1}{2}P \cdot \frac{\partial}{\partial t}\mathcal{E}$, and using the chain rule leads to

$$-\mathcal{E} \cdot \frac{\partial D}{\partial t} = -\frac{1}{2}\mathcal{E} \cdot \frac{\partial}{\partial t}\left(\epsilon_0 \mathcal{E}\right) - \frac{1}{2}\mathcal{E} \cdot \frac{\partial}{\partial t}P - \frac{1}{2}\frac{\partial}{\partial t}\mathcal{E} \cdot \left(\epsilon_0 \mathcal{E}\right) - \frac{1}{2}\frac{\partial}{\partial t}\left(\mathcal{E} \cdot P\right)$$
$$-\frac{1}{2}\mathcal{E} \cdot \frac{\partial}{\partial t}P + \frac{1}{2}P \cdot \frac{\partial}{\partial t}\mathcal{E}, \tag{2.60}$$

Grouping the first two, middle two, and last two terms of the right-hand side reformulates this relation as

$$-\mathcal{E} \cdot \frac{\partial D}{\partial t} = -\frac{1}{2}\mathcal{E} \cdot \frac{\partial}{\partial t}(\epsilon_0 \mathcal{E} + P) - \frac{1}{2}\frac{\partial}{\partial t}\left[\mathcal{E} \cdot (\epsilon_0 \mathcal{E} + P)\right]$$
$$-\frac{1}{2}\left(\mathcal{E} \cdot \frac{\partial}{\partial t}P - P \cdot \frac{\partial}{\partial t}\mathcal{E}\right), \tag{2.61}$$

which, using again $D = \epsilon_0 \mathcal{E} + P$, becomes

$$-\mathcal{E} \cdot \frac{\partial D}{\partial t} = -\frac{1}{2}\mathcal{E} \cdot \frac{\partial}{\partial t}D - \frac{1}{2}\frac{\partial}{\partial t}\mathcal{E} \cdot D - \frac{1}{2}\left(\mathcal{E} \cdot \frac{\partial}{\partial t}P - P \cdot \frac{\partial}{\partial t}\mathcal{E}\right). \tag{2.62}$$

Finally, grouping the first two terms of the right-hand side of this relation yields

$$-\mathcal{E} \cdot \frac{\partial D}{\partial t} = -\frac{1}{2}\frac{\partial}{\partial t}(\mathcal{E} \cdot D) - \frac{1}{2}\left(\mathcal{E} \cdot \frac{\partial}{\partial t}P - P \cdot \frac{\partial}{\partial t}\mathcal{E}\right). \tag{2.63}$$

Similarly, the term $-H \cdot \frac{\partial B}{\partial t}$ in (2.57) becomes

$$-H \cdot \frac{\partial B}{\partial t} = -\frac{1}{2}\frac{\partial}{\partial t}(H \cdot B) - \frac{\mu_0}{2}\left(H \cdot \frac{\partial}{\partial t}M - M \cdot \frac{\partial}{\partial t}H\right). \tag{2.64}$$

Substituting now (2.63) and (2.64) into (2.57) finally yields the bianisotropic Poynting theorem:

$$\frac{\partial w}{\partial t} + \nabla \cdot S = -I_J - I_K - I_P - I_M, \tag{2.65}$$

where w is the energy density, S is the Poynting vector, and I_J and I_K are the impressed source power densities, and I_P and I_M are the induced polarization power densities, respectively, which are defined by

$$w = \frac{1}{2}(\mathcal{E} \cdot D + H \cdot B), \tag{2.66a}$$

$$I_J = \mathcal{E} \cdot J, \tag{2.66b}$$

$$I_K = H \cdot K, \tag{2.66c}$$

$$I_P = \frac{1}{2}\left(\mathcal{E} \cdot \frac{\partial}{\partial t}P - P \cdot \frac{\partial}{\partial t}\mathcal{E}\right), \tag{2.66d}$$

$$I_M = \frac{\mu_0}{2}\left(\mathcal{H} \cdot \frac{\partial}{\partial t}\mathcal{M} - \mathcal{M} \cdot \frac{\partial}{\partial t}\mathcal{H}\right). \tag{2.66e}$$

It is generally more convenient to consider the time-average version of (2.65), which reads

$$\left\langle \frac{\partial w}{\partial t}\right\rangle + \nabla \cdot \langle \mathbf{S}\rangle = -\langle I_J\rangle - \langle I_K\rangle - \langle I_P\rangle - \langle I_M\rangle, \tag{2.67}$$

where $\langle \cdot \rangle$ denotes the time-average operation. In the case of time-harmonic fields, this equation can be further manipulated as follows. First, it is straightforward to show that $\langle \frac{\partial w}{\partial t}\rangle = 0$ and $\langle \mathbf{S}\rangle = \frac{1}{2}\mathrm{Re}(\mathcal{E} \times \mathcal{H}^*)$. Second, the terms $\langle I_J\rangle$ and $\langle I_K\rangle$ may be expressed in terms of the electric and magnetic susceptibility tensors by using $\mathbf{J} = \overline{\overline{\sigma}}_e \cdot \mathbf{E}$ and $\mathbf{K} = \overline{\overline{\sigma}}_m \cdot \mathbf{H}$, where the electric and magnetic conductivity tensors are related to the susceptibility tensors as

$$\overline{\overline{\sigma}}_e = -\omega\epsilon_0\mathrm{Im}(\overline{\overline{\chi}}_{ee}) = \frac{j\omega\epsilon_0}{2}(\overline{\overline{\chi}}_{ee} - \overline{\overline{\chi}}_{ee}^*), \tag{2.68a}$$

$$\overline{\overline{\sigma}}_m = -\omega\mu_0\mathrm{Im}(\overline{\overline{\chi}}_{mm}) = \frac{j\omega\mu_0}{2}(\overline{\overline{\chi}}_{mm} - \overline{\overline{\chi}}_{mm}^*), \tag{2.68b}$$

which leads, after replacing the instantaneous field vectors by their phaser counterparts, to

$$\langle I_J\rangle = \frac{1}{2}\mathrm{Re}(\mathbf{E}^* \cdot \mathbf{J}) = \frac{1}{2}\mathrm{Re}(\mathbf{E}^* \cdot \overline{\overline{\sigma}}_e \cdot \mathbf{E}) = \frac{1}{4}\mathrm{Re}\left[j\omega\epsilon_0\mathbf{E}^* \cdot (\overline{\overline{\chi}}_{ee} - \overline{\overline{\chi}}_{ee}^*) \cdot \mathbf{E}\right], \tag{2.69a}$$

$$\langle I_K\rangle = \frac{1}{2}\mathrm{Re}(\mathbf{H}^* \cdot \mathbf{K}) = \frac{1}{2}\mathrm{Re}(\mathbf{H}^* \cdot \overline{\overline{\sigma}}_m \cdot \mathbf{H})$$
$$= \frac{1}{4}\mathrm{Re}\left[j\omega\mu_0\mathbf{H}^* \cdot (\overline{\overline{\chi}}_{mm} - \overline{\overline{\chi}}_{mm}^*) \cdot \mathbf{H}\right]. \tag{2.69b}$$

Third, we write the terms $\langle I_P\rangle$ and $\langle I_M\rangle$ in (2.67) in the frequency domain, which transforms the time-derivatives in 2.66d and 2.66e according to $\frac{\partial}{\partial t} \to j\omega$, so that

$$\langle I_P\rangle = \frac{1}{4}\mathrm{Re}\left[\mathbf{E}^* \cdot (j\omega\mathbf{P}) - \mathbf{P}^* \cdot (j\omega\mathbf{E})\right] = \frac{1}{4}\mathrm{Re}\left[j\omega(\mathbf{E}^* \cdot \mathbf{P} - \mathbf{P}^* \cdot \mathbf{E})\right], \tag{2.70a}$$

$$\langle I_M\rangle = \frac{\mu_0}{4}\mathrm{Re}\left[\mathbf{H}^* \cdot (j\omega\mathbf{M}) - \mathbf{M}^* \cdot (j\omega\mathbf{H})\right] = \frac{1}{4}\mathrm{Re}\left[j\omega\mu_0(\mathbf{H}^* \cdot \mathbf{M} - \mathbf{M}^* \cdot \mathbf{H})\right]. \tag{2.70b}$$

Substituting $\mathbf{P}$ and $\mathbf{M}$ in (2.70) by their constitutive definitions in (2.54) transforms (2.70) into

$$\langle I_P\rangle = \frac{1}{4}\mathrm{Re}\left[j\omega\epsilon_0(\mathbf{E}^* \cdot (\overline{\overline{\chi}}_{ee} \cdot \mathbf{E} + \eta_0\overline{\overline{\chi}}_{em} \cdot \mathbf{H}) - \mathbf{E} \cdot (\overline{\overline{\chi}}_{ee} \cdot \mathbf{E} + \eta_0\overline{\overline{\chi}}_{em} \cdot \mathbf{H})^*)\right], \tag{2.71a}$$

$$\langle I_M\rangle = \frac{1}{4}\mathrm{Re}\left[j\omega\mu_0(\mathbf{H}^* \cdot (\overline{\overline{\chi}}_{mm} \cdot \mathbf{H} + \overline{\overline{\chi}}_{me} \cdot \mathbf{E}/\eta_0)\right.$$
$$\left. -\mathbf{H} \cdot (\overline{\overline{\chi}}_{mm} \cdot \mathbf{H} + \overline{\overline{\chi}}_{me} \cdot \mathbf{E}/\eta_0)^*)\right]. \tag{2.71b}$$

Finally, rearranging and simplifying the terms in (2.71) leads to

$$\langle I_{\mathrm{P}} \rangle = \frac{1}{4} \mathrm{Re} \left[j\omega\epsilon_0 (\mathbf{E}^* \cdot (\overline{\overline{\chi}}_{ee} - \overline{\overline{\chi}}^\dagger_{ee}) \cdot \mathbf{E} + 2\eta_0 \mathbf{E}^* \cdot \overline{\overline{\chi}}_{em} \cdot \mathbf{H}) \right], \qquad (2.72a)$$

$$\langle I_{\mathrm{M}} \rangle = \frac{1}{4} \mathrm{Re} \left[j\omega\mu_0 (\mathbf{H}^* \cdot (\overline{\overline{\chi}}_{mm} - \overline{\overline{\chi}}^\dagger_{mm}) \cdot \mathbf{H} - 2\mathbf{E}^* \cdot \overline{\overline{\chi}}^\dagger_{me} \cdot \mathbf{H}/\eta_0) \right], \qquad (2.72b)$$

where the superscript † corresponds to the transpose conjugate operation. The final expression of the time-average bianisotropic Poynting theorem for time-harmonic fields is then given by

$$\nabla \cdot \langle \mathbf{S} \rangle = - \left\langle I_{\mathrm{J}} \right\rangle - \langle I_{\mathrm{K}} \rangle - \langle I_{\mathrm{P}} \rangle - \langle I_{\mathrm{M}} \rangle, \qquad (2.73)$$

where $\left\langle I_{\mathrm{J}} \right\rangle$, $\langle I_{\mathrm{K}} \rangle$ and $\langle I_{\mathrm{P}} \rangle$, $\langle I_{\mathrm{M}} \rangle$ are, respectively, provided by (2.69) and (2.72).

Integrating (2.73) over a volume V defined by the surface S, and applying the divergence theorem to the resulting left-hand side transforms this relation into

$$\iint_S \langle \mathbf{S} \rangle \cdot \mathrm{d}\mathbf{S} = - \iiint_V \left(\langle I_{\mathrm{J}} \rangle + \langle I_{\mathrm{K}} \rangle + \langle I_{\mathrm{P}} \rangle + \langle I_{\mathrm{M}} \rangle \right) \mathrm{d}V, \qquad (2.74)$$

which indicates that the amount of loss or gain that an electromagnetic wave experiences in a given volume surrounding a medium is related to the amount of energy passing across the surface delimiting this volume.

If the medium is perfectly lossless and gainless, the amounts of electromagnetic energy entering and exiting the medium are equal, so that $\iint_S \langle \mathbf{S} \rangle \cdot \mathrm{d}\mathbf{S} = 0$ in (2.74), and therefore $\nabla \cdot \langle \mathbf{S} \rangle = 0$ in (2.73). In the case of a lossy medium, there is less energy leaving than entering the volume, corresponding to $\nabla \cdot \langle \mathbf{S} \rangle < 0$, and vice versa for a gain medium, i.e. $\nabla \cdot \langle \mathbf{S} \rangle > 0$. Note that the fact that $\nabla \cdot \langle \mathbf{S} \rangle = 0$ is a necessary but not sufficient condition for the medium to be gainless and lossless, since gain–loss compensation may occur between the terms on the right-hand side of (2.73).

Substituting (2.69) and (2.72) into (2.73) provides the following alternative form of the Poynting theorem:

$$\nabla \cdot \langle \mathbf{S} \rangle = -\frac{1}{4} \mathrm{Re} \left[j\omega(\epsilon_0 \mathbf{E}^* \cdot (2\overline{\overline{\chi}}_{ee} - \overline{\overline{\chi}}^*_{ee} - \overline{\overline{\chi}}^\dagger_{ee}) \cdot \mathbf{E} \right.$$
$$\left. + \mu_0 \mathbf{H}^* \cdot (2\overline{\overline{\chi}}_{mm} - \overline{\overline{\chi}}^*_{mm} - \overline{\overline{\chi}}^\dagger_{mm}) \cdot \mathbf{H} + 2k_0 \mathbf{E}^* \cdot (\overline{\overline{\chi}}_{me} - \overline{\overline{\chi}}^\dagger_{em}) \cdot \mathbf{H}) \right]. \qquad (2.75)$$

In the absence of impressed surface currents, i.e. $\left\langle I_{\mathrm{J}} \right\rangle = \langle I_{\mathrm{K}} \rangle = 0$, then (2.75) reduces to

$$\nabla \cdot \langle \mathbf{S} \rangle = -\frac{1}{4} \mathrm{Re} \left[j\omega(\epsilon_0 \mathbf{E}^* \cdot (\overline{\overline{\chi}}_{ee} - \overline{\overline{\chi}}^\dagger_{ee}) \cdot \mathbf{E} \right.$$
$$\left. + \mu_0 \mathbf{H}^* \cdot (\overline{\overline{\chi}}_{mm} - \overline{\overline{\chi}}^\dagger_{mm}) \cdot \mathbf{H} + 2k_0 \mathbf{E}^* \cdot (\overline{\overline{\chi}}_{me} - \overline{\overline{\chi}}^\dagger_{em}) \cdot \mathbf{H}) \right]. \qquad (2.76)$$

2.6 Energy Conservation in Lossless–Gainless Systems

From a practical perspective, it is often desirable to implement gainless and lossless systems. This specification requires certain conditions to be satisfied, which may be expressed either in terms of susceptibilities or scattering parameters. We shall next derive and discuss these conditions.

The conditions in terms of the susceptibilities may be directly deduced from the Poynting theorem. As shown in Section 2.5, a medium is gainless and lossless if Eq. (2.76) is zero for any field value, which implies that

$$\overline{\overline{\chi}}_{ee} = \overline{\overline{\chi}}^{\dagger}_{ee}, \quad \overline{\overline{\chi}}_{mm} = \overline{\overline{\chi}}^{\dagger}_{mm}, \quad \overline{\overline{\chi}}_{me} = \overline{\overline{\chi}}^{\dagger}_{em}. \tag{2.77}$$

Since these conditions are similar to the reciprocity conditions (2.51), it is interesting to see under which circumstances a medium satisfies (2.77) while being either reciprocal or nonreciprocal. For this purpose, Table 2.2 compares the different possible cases.

This table reveals that a gainless, lossless, and reciprocal medium has purely real electric and magnetic susceptibility tensors and purely imaginary magnetoelectric susceptibilities. Interestingly, we also see that it is possible for a medium to be gainless, lossless, and nonreciprocal.

We now express the corresponding relationships in terms of scattering parameters. For this purpose, consider a uniform bianisotropic slab sandwiched between two different media and illuminated by the normally incident waves $U_1^{(+)}$ and $U_2^{(-)}$ and scattering the waves $U_1^{(-)}$ and $U_2^{(+)}$, where U represents the electric field amplitude of the waves, as depicted in Figure 2.4. Here, we assume that the waves are all identically polarized.

The $U_i^{\pm}(i = 1, 2)$ quantities are connected to each other via scattering parameters, S_{ij}, as

$$\begin{bmatrix} U_2^{(+)} \\ U_1^{(-)} \end{bmatrix} = \begin{bmatrix} S_{21} & S_{22} \\ S_{11} & S_{12} \end{bmatrix} \begin{bmatrix} U_1^{(+)} \\ U_2^{(-)} \end{bmatrix}. \tag{2.78}$$

Table 2.2 Conditions for a medium to be gainless and lossless in addition to being either reciprocal or nonreciprocal.

	$\overline{\overline{\chi}}_{ee}$	$\overline{\overline{\chi}}_{mm}$	$\overline{\overline{\chi}}_{em}$ and $\overline{\overline{\chi}}_{me}$
Reciprocal	$\mathrm{Im}(\overline{\overline{\chi}}_{ee}) = 0$ $\overline{\overline{\chi}}_{ee} = \overline{\overline{\chi}}^{T}_{ee}$	$\mathrm{Im}(\overline{\overline{\chi}}_{mm}) = 0$ $\overline{\overline{\chi}}_{mm} = \overline{\overline{\chi}}^{T}_{mm}$	$\mathrm{Re}(\overline{\overline{\chi}}_{me,em}) = 0$ $\overline{\overline{\chi}}_{me} = -\overline{\overline{\chi}}^{T}_{em}$
Nonreciprocal	$\overline{\overline{\chi}}_{ee} \neq \overline{\overline{\chi}}^{T}_{ee}$ $\overline{\overline{\chi}}_{ee} = \overline{\overline{\chi}}^{\dagger}_{ee}$	$\overline{\overline{\chi}}_{mm} \neq \overline{\overline{\chi}}^{T}_{mm}$ $\overline{\overline{\chi}}_{mm} = \overline{\overline{\chi}}^{\dagger}_{mm}$	$\mathrm{Re}(\overline{\overline{\chi}}_{me,em}) \neq 0$ $\overline{\overline{\chi}}_{me} = \overline{\overline{\chi}}^{\dagger}_{em}$

Figure 2.4 Normal scattering by a bianisotropic slab sandwiched between two media.

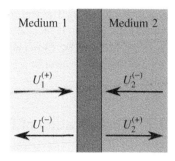

Since the slab is gainless and lossless, conservation of power requires that the scattered power equals the incident power, i.e.

$$\frac{|U_1^{(+)}|^2}{\eta_1} + \frac{|U_2^{(-)}|^2}{\eta_2} = \frac{|U_1^{(-)}|^2}{\eta_1} + \frac{|U_2^{(+)}|^2}{\eta_2}, \tag{2.79}$$

where η_1 and η_2 are the impedances of medium 1 and 2, respectively, and are assumed to be real quantities. Substituting (2.78) into (2.79), and considering that $|U|^2 = UU^*$, yields

$$|U_1^{(+)}|^2 \left(\frac{1}{\eta_1} - \frac{|S_{11}|^2}{\eta_1} - \frac{|S_{21}|^2}{\eta_2} \right) + |U_2^{(-)}|^2 \left(\frac{1}{\eta_2} - \frac{|S_{22}|^2}{\eta_2} - \frac{|S_{12}|^2}{\eta_1} \right)$$
$$= U_1^{(+)} U_2^{*(-)} \left(\frac{S_{11} S_{12}^*}{\eta_1} + \frac{S_{22}^* S_{21}}{\eta_2} \right) + U_1^{*(+)} U_2^{(-)} \left(\frac{S_{11}^* S_{12}}{\eta_1} + \frac{S_{22} S_{21}^*}{\eta_2} \right). \tag{2.80}$$

Since this equality holds for any field value, all the terms in brackets must vanish. We have thus

$$|S_{11}|^2 + |S_{21}|^2 \frac{\eta_1}{\eta_2} = 1 \quad \text{and} \quad |S_{22}|^2 + |S_{12}|^2 \frac{\eta_2}{\eta_1} = 1, \tag{2.81}$$

which indicates how the incident power distributes in terms of the reflected and transmitted power, and

$$\frac{S_{11} S_{12}^*}{\eta_1} + \frac{S_{22}^* S_{21}}{\eta_2} = 0 \quad \text{and} \quad \frac{S_{11}^* S_{12}}{\eta_1} + \frac{S_{22} S_{21}^*}{\eta_2} = 0, \tag{2.82}$$

which implies that the phases of the scattered waves are related to each other.

For a reciprocal slab with the same media at both sides, i.e. $\eta_1 = \eta_2$, these expressions reduce to

$$|S_{11}|^2 + |S_{21}|^2 = 1 \quad \text{and} \quad |S_{22}|^2 + |S_{12}|^2 = 1, \tag{2.83}$$

and

$$\frac{t^*}{t} = -\frac{S_{22}^*}{S_{11}} \quad \text{and} \quad |S_{11}|^2 = |S_{22}|^2, \tag{2.84}$$

where $t = S_{12} = S_{21}$ by reciprocity. Interestingly, if the slab is gainless, lossless, reciprocal, and isotropic, as for instance a thin beam splitting blade, then $S_{11} = S_{22} = r$ and Eq. (2.84) tells us that the argument of t must necessarily be in quadrature with respect to r.

2.7 Classification of Bianisotropic Media

It is sometimes convenient to classify bianisotropic media in terms of their tensorial constitutive parameters. For this purpose, we split the susceptibility tensors $\overline{\overline{\chi}}_{em}$ and $\overline{\overline{\chi}}_{me}$ into a tensor, $\overline{\overline{v}}$, related to nonreciprocity, and a tensor, $\overline{\overline{\kappa}}$, related to chirality [148], as

$$\overline{\overline{\chi}}_{em} = \overline{\overline{v}} - j\overline{\overline{\kappa}} \quad \text{and} \quad \overline{\overline{\chi}}_{me} = (\overline{\overline{v}} + j\overline{\overline{\kappa}})^{\mathrm{T}}. \tag{2.85}$$

Table 2.3 classifies the reciprocal ($\overline{\overline{v}} = 0$ and $\overline{\overline{\kappa}} \neq 0$) and nonreciprocal ($\overline{\overline{v}} \neq 0$ and $\overline{\overline{\kappa}} = 0$) bianisotropic media depending on whether the tensors $\overline{\overline{v}}$ or $\overline{\overline{\kappa}}$ are diagonal or not, and symmetric or antisymmetric, respectively. This classification relies on the fact that an arbitrary tensor $\overline{\overline{M}}$, which may represent either $\overline{\overline{\kappa}}$ (for a reciprocal

Table 2.3 Classification of bianisotropic media [148].

Type	Parameters	Medium
Reciprocal	$\overline{\overline{D}} = 0, \overline{\overline{S}} = 0, \overline{\overline{A}} \neq 0$	Omega
$\overline{\overline{v}} = 0$	$\overline{\overline{D}} = 0, \overline{\overline{S}} \neq 0, \overline{\overline{A}} = 0$	Pseudochiral
$\overline{\overline{\kappa}} \neq 0$	$\overline{\overline{D}} = 0, \overline{\overline{S}} \neq 0, \overline{\overline{A}} \neq 0$	Pseudochiral omega
	$\overline{\overline{D}} \neq 0, \overline{\overline{S}} = 0, \overline{\overline{A}} = 0$	Pasteur (or biisotropic)
	$\overline{\overline{D}} \neq 0, \overline{\overline{S}} = 0, \overline{\overline{A}} \neq 0$	Chiral omega
	$\overline{\overline{D}} \neq 0, \overline{\overline{S}} \neq 0, \overline{\overline{A}} = 0$	Anisotropic chiral
	$\overline{\overline{D}} \neq 0, \overline{\overline{S}} \neq 0, \overline{\overline{A}} \neq 0$	General reciprocal
Nonreciprocal	$\overline{\overline{D}} = 0, \overline{\overline{S}} = 0, \overline{\overline{A}} \neq 0$	Moving
$\overline{\overline{v}} \neq 0$	$\overline{\overline{D}} = 0, \overline{\overline{S}} \neq 0, \overline{\overline{A}} = 0$	Pseudo Tellegen
$\overline{\overline{\kappa}} = 0$	$\overline{\overline{D}} = 0, \overline{\overline{S}} \neq 0, \overline{\overline{A}} \neq 0$	Moving pseudo Tellegen
	$\overline{\overline{D}} \neq 0, \overline{\overline{S}} = 0, \overline{\overline{A}} = 0$	Tellegen
	$\overline{\overline{D}} \neq 0, \overline{\overline{S}} = 0, \overline{\overline{A}} \neq 0$	Moving Tellegen
	$\overline{\overline{D}} \neq 0, \overline{\overline{S}} \neq 0, \overline{\overline{A}} = 0$	Anisotropic Tellegen
	$\overline{\overline{D}} \neq 0, \overline{\overline{S}} \neq 0, \overline{\overline{A}} \neq 0$	Nonreciprocal nonchiral

medium) or $\overline{\overline{v}}$ (for a nonreciprocal medium), can be decomposed as

$$\overline{\overline{M}} = \overline{\overline{D}} + \overline{\overline{N}}, \tag{2.86}$$

where $\overline{\overline{D}}$ is a diagonal tensor and $\overline{\overline{N}}$ is a traceless tensor that can be further split into a symmetric part, $\overline{\overline{S}}$, and an antisymmetric part, $\overline{\overline{A}}$, where

$$\overline{\overline{S}} = \frac{\overline{\overline{N}} + \overline{\overline{N}}^{T}}{2} \quad \text{and} \quad \overline{\overline{A}} = \frac{\overline{\overline{N}} - \overline{\overline{N}}^{T}}{2}, \tag{2.87}$$

such that (2.86) becomes

$$\overline{\overline{M}} = \overline{\overline{D}} + \overline{\overline{S}} + \overline{\overline{A}}. \tag{2.88}$$

3

Metasurface Modeling

In Chapter 1, we have discussed the concept of metasurfaces from a very general perspective. We have then presented fundamental properties of materials in Chapter 2. In this chapter, we develop the mathematical tools required to model metasurfaces. We present three modeling techniques that are based on different quantities: impedances and admittances, effective polarizabilities, and effective susceptibilities. While these modeling techniques approach the modeling of metasurfaces from different perspectives, they are all based on the same two fundamental principles: homogeneity and zero thickness. We will first discuss these two fundamental concepts, then demonstrate their physical validity, and finally show how they can be applied to metasurfaces. Accordingly, Section 3.1 shows that a metasurface may be modeled as an *homogeneous* medium despite the fact that it generally consists of a periodic array of scattering particles. Then, Section 3.2 shows that a metasurface may be modeled as a *zero-thickness* sheet, since it is practically much thinner than the wavelength(s) of operation. Finally, Section 3.3 introduces and compares the three metasurface modeling techniques mentioned above.

3.1 Effective Homogeneity

3.1.1 The Homogeneity Paradox

One of the most fundamental properties of a metasurface is that it *effectively* behaves as a homogeneous electromagnetic medium despite the fact that a metasurface is typically made of a periodic or quasiperiodic array of scattering particles. The fact that a metasurface is "seen" by electromagnetic waves as an homogeneous medium may a priori seem in contradiction with the fact that an array of scattering particles may inherently be described as a spatially varying refractive index function. The goal of this section is to clarify this point. For that

Electromagnetic Metasurfaces: Theory and Applications, First Edition.
Karim Achouri and Christophe Caloz.
© 2021 The Institute of Electrical and Electronics Engineers, Inc.
Published 2021 by John Wiley & Sons, Inc.

Figure 3.1 1D periodic array of period L.

purpose, we will first review the theory of periodic structures and then investigate in which situations an array may be modeled as an homogenized medium.

3.1.2 Theory of Periodic Structures

Let us consider the one-dimensional (1D) periodic structure in Figure 3.1, which consists of the repetition of a unit cell of length L and is composed of two sections with respective refractive indices n_a and n_b.

The Bloch–Floquet theorem stipulates that a wave, $\Psi(x)$, interacting with this structure must satisfy the translation symmetry relation [67]

$$\frac{\Psi(x+L)}{\Psi(x)} = \frac{\Psi(x+2L)}{\Psi(x+L)} = \cdots = \frac{\Psi(x+pL)}{\Psi[x+(p-1)L]} = C, \tag{3.1}$$

where p is an integer spanning $]-\infty, +\infty[$ and C is a complex constant of unit magnitude, because the fields at a given point of adjacent cells can differ only by a phase factor, i.e.

$$C = e^{-jk_x L}, \tag{3.2}$$

where k_x is the complex propagation constant along x. Equation (3.1) may be manipulated as

$$\Psi(x+pL) = C\Psi[x+(p-1)L] = C\{C\Psi[x+(p-2)L]\} = \cdots = C^p\Psi(x), \tag{3.3}$$

which suggests that $\Psi(x)$ is related to a periodic function of the form

$$A(x) = e^{jk_x x}\Psi(x). \tag{3.4}$$

Indeed, making use of (3.2) and (3.3), we have that $A(x+L)$ becomes

$$A(x+L) = e^{jk_x(x+L)}\Psi(x+L) = e^{jk_x(x+L)}C\Psi(x) = e^{jk_x x}\Psi(x) = A(x), \tag{3.5}$$

where $A(x)$ is the sought-after periodic function of period L. As any periodic function, $A(x)$ may be expressed as a Fourier series expansion, i.e.

$$A(x) = \sum_{m=-\infty}^{+\infty} A_m e^{-j\frac{2m\pi}{L}x}. \tag{3.6}$$

Inserting this relation into (3.4), we obtain

$$\Psi(x) = \sum_{m=-\infty}^{+\infty} A_m e^{-jk_{x,m}x}, \quad \text{with} \quad k_{x,m} = k_x + \frac{2m\pi}{L}, \tag{3.7}$$

which corresponds to a superposition of plane waves with propagation constant $k_{x,m}$ related to the period L of the structure, and constitutes the final form of the Bloch–Floquet theorem.

3.1.3 Scattering from Gratings

Equation (3.7) pertains to a 1D structure, but may naturally be extended to the cases of 2D and 3D structures. A situation of particular interest is that depicted in Figure 3.2, which consists of a subwavelengthly thin periodic grating lying in between two half spaces of lossless refractive indices n_1 and n_2, respectively.

A plane wave, $\Psi_i(x, z)$, obliquely impinges on the grating, which scatters the reflected and transmitted waves

$$\Psi_r(x, z) = \sum_{m=-\infty}^{+\infty} A_{r,m} e^{-jk_{x,m}x} \quad \text{and} \quad \Psi_t(x, z) = \sum_{m=-\infty}^{+\infty} A_{t,m} e^{-jk_{x,m}x}, \tag{3.8}$$

where $k_{x,m}$ is given in (3.7), and is the same for the reflected and transmitted waves from continuity of the tangential field components (phase matching) across the grating. The plane waves in (3.8) are conventionally referred to as diffraction orders, of order m, and propagate in different directions of space, as illustrated in Figure 3.2 for orders $m = \{-1, 0, +1\}$. The longitudinal propagation constants of the reflected and transmitted diffraction orders may be respectively expressed as

$$k_{r,z} = -\sqrt{k_1^2 - k_{x,m}^2} \quad \text{and} \quad k_{t,z} = \sqrt{k_2^2 - k_{x,m}^2}, \tag{3.9}$$

Figure 3.2 Scattering by a subwavelengthly thick grating of period L. The medium above the grating has a refractive index n_2, while the medium below has a refractive index n_1.

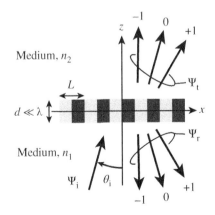

Metasurface   Homogenized medium   Zerothickness model

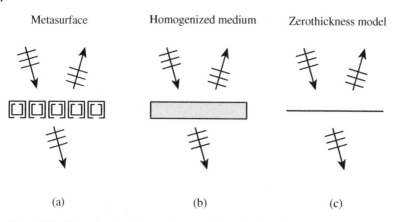

(a)   (b)   (c)

Figure 3.3 Three level of metasurface modeling. (a) Generic metasurface made of a subwavelength periodic array of split-ring resonators. (b) Slab with 3D homogenized effective material parameters corresponding to the metasurface on in (a). (c) Zero-thickness sheet model of (b) with effective 2D material parameters.

where $k_1 = k_0 n_1$ and $k_2 = k_0 n_2$, and where k_x corresponds to the tangential wavenumber of the incident wave, Ψ_i. These expressions may be used to determine whether the diffraction orders are propagating ($k_{z,m} \in \mathbb{R}$) or evanescent ($k_{z,m} \in \mathbb{I}$) along the z-direction.

3.1.4 Homogenization

We may now come back to the problem of metasurface homogenization. What is the condition for a metasurface to be homogenizable into a thin uniform slab characterized by effective material parameters, as illustrated in the passage from Figure 3.3a to b? By definition, a thin uniform slab does not produce any diffraction order, and hence simply refracts light according to Snells law.

The condition of homogeneity is that only the waves corresponding to $m = 0$ in (3.8) propagate to the far-field of the metasurface, while all other waves remain confined to it as surface waves. This implies a specific relationship between the period of the array and the wavelength of the incident wave. This relationship can be found by inserting the definition of $k_{x,m}$ in (3.7) into (3.9), which gives[1]

$$\sqrt{k^2 - k_{x,m}^2} = \sqrt{k^2 - \left(k_x + \frac{2m\pi}{L}\right)^2},\tag{3.10}$$

---

1 Here, we only consider the case of diffraction orders along the x-direction but identical considerations apply for diffraction orders along the y-direction.

where k represents either k_1 or k_2. In order to avoid the nonzero diffraction orders, the term under the square root must be negative for $m \neq 0$, which occurs if

$$\pm k < k_x + \frac{2m\pi}{L}. \tag{3.11}$$

Thus, suppressing the diffraction orders corresponding to $m \neq 0$ requires restricting L to be small enough so that (3.11) is satisfied for $|m| = 1$, which automatically ensures that (3.11) is also satisfied for $|m| > 1$, i.e.

$$L < \left| \frac{2\pi}{k - k_x} \right| \tag{3.12}$$

Let us now consider that the incident wave, Ψ_i, is impinging on the periodic structure from medium 1 with an incidence angle θ_i, as shown in Figure 3.2, so that $k_x = k_1 \sin \theta_i$ in (3.11). Equation (3.12) provides then the upper bound

$$L < \frac{\lambda_0}{n_{1,2} + n_1 |\sin \theta_i|}, \tag{3.13}$$

where $n_{1,2}$ stands for n_1 for the reflected diffraction orders, and for n_2 for the transmitted diffraction orders. Equation (3.13) shows that the condition on L is getting increasingly restrictive as the angle θ_i increases, reaching $L < \lambda_0/(n_{1,2} + n_1)$ at $\theta_i = 90°$, where the incident plane wave is grazing along the structure. In this case, we have

$$L < \min \left(\frac{\lambda_0}{2n_1}, \frac{\lambda_0}{n_2 + n_1} \right). \tag{3.14}$$

A convenient rule of thumb is to approximate (3.14) by the relation

$$L < \frac{\lambda_{\text{av}}}{2}, \tag{3.15}$$

where $\lambda_{\text{av}} = 2\lambda_0/(n_1 + n_2)$. Thus, a thin periodic structure is essentially homogenizable when its period is less than half the average wavelength. Note that if the incident wave is normally incident on the array ($\theta_i = 0$), then the condition (3.15) is relaxed to $L < \lambda_{\text{av}}$. This indicates that metasurface homogenization is easier to achieve close to normal incidence than to grazing angles.

3.2 Effective Zero Thickness

We have just seen that an electromagnetically thin periodic array of subwavelength period ($L < \lambda_{\text{av}}/2$), such as that in Figure 3.3a, may be homogenized into a thin slab exhibiting effective material parameters, as the one in Figure 3.3b.

In order to obtain its effective material parameters, we need to develop an appropriate theoretical model. Since it is electromagnetically thin ($d/\lambda \to 0$), spatial variations of the fields across it are negligible. Therefore, it may be modeled as a zero-thickness sheet, as shown in Figure 3.3c. Mathematically, this can be expressed, using the notation in Figure 3.2, as

$$\frac{\partial}{\partial z}\Psi(x,z) \approx 0 \quad \text{for} \quad z = [-d/2, d/2], \tag{3.16}$$

where $\Psi(x,z)$ represents the total field within the metasurface. This approximation is suitable in most cases, as will be shown, but also has limitations, which we shall discuss at several occasions throughout the book.

The main reason for considering a metasurface as a zero-thickness sheet is the dramatic related simplifications. Indeed, modeling the interactions of a slab with an electromagnetic wave and retrieving its effective material parameters would require: (i) satisfying the boundary conditions at the *two* interfaces of the slab, (ii) accounting for multiple wave scattering within the slab, and (iii) dealing with 3D material parameters. In contrast, a zero-thickness sheet possesses only *one* interface, which implies that (i) the boundary conditions must be satisfied only once, (ii) there is no scattering within the sheet, and (iii) we are manipulating only 2D material parameters. Overall, modeling a metasurface as a zero-thickness sheet is practically very effective and, generally, only comes at the cost of negligible errors. Such errors are, for instance, discussed in [37] where the modeling of a nonzero-thickness metasurface is presented.

From now on, we will thus consider that metasurfaces may be modeled as zero-thickness sheets, as in Figure 3.3c, and we will therefore deal with 2D material parameters. One may naturally wonder how 2D material parameters practically apply to physical structures whose thickness is really nonzero. It turns out that a simple relationship exists between 3D and 2D material parameters as long as the physical structure under consideration satisfies the condition (3.16). We next derive this relationship from the conventional boundary conditions.

For simplicity, and without loss of generality, let us consider an isotropic and uniform metasurface, characterized by the electric scalar susceptibility, χ_{ee}. The corresponding Maxwell–Ampère equation, assuming time-harmonic fields with dependence $e^{+j\omega t}$, reads

$$\nabla \times \mathbf{H} = j\omega\epsilon_0(1 + \chi_{ee})\mathbf{E}, \tag{3.17}$$

where $1 + \chi_{ee} = \epsilon_r$ is the metasurface relative permittivity. Let us now consider the two cases where (i) the metasurface is of exactly zero thickness, for which $\chi_{ee} = \chi_{2D}\delta(z)$ with $\delta(z)$ being the Dirac delta distribution, and (ii) the metasurface has a nonzero subwavelength thickness d, for which $\chi_{ee} = \chi_{3D}\Pi(z/d)$ with $\Pi(z)$ being the rectangular function. These two cases are depicted in Figure 3.4a and b, respectively, where the metasurface is assumed in both cases to be of infinite extent in the *xy*-plane.

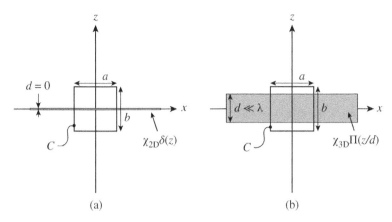

Figure 3.4 Comparison between the surface and volume models of a metasurface. (a) Zero-thickness sheet model. (b) Slab volume model.

Integrating (3.17) over the surface $S = ab$ in Figure 3.4a, and applying the Stokes theorem gives

$$\oint_C \mathbf{H} \cdot d\mathbf{l} = j\omega\epsilon_0 \iint_S \left[1 + \chi_{2D}\delta(z)\right] \mathbf{E} \cdot d\mathbf{S}, \tag{3.18}$$

where C is the contour delimiting S. Letting $b \to 0$ reduces (3.18) to

$$(H_x^+ - H_x^-)a \approx j\omega\epsilon_0 \left(b + \chi_{2D}\right) E_y a, \tag{3.19}$$

where it has been assumed that a and b are very small compared to the wavelength is dimensionless. Similar relations theorem to (3.17) for the case of Figure 3.4b gives

$$\oint_C \mathbf{H} \cdot d\mathbf{l} = j\omega\epsilon_0 \iint_S \left[1 + \chi_{3D}\Pi(z/d)\right] \mathbf{E} \cdot d\mathbf{S}, \tag{3.20}$$

which reduces to

$$(H_x^+ - H_x^-)a \approx j\omega\epsilon_0 \left(b + \chi_{3D}d\right) E_y a. \tag{3.21}$$

Comparing the expressions in (3.19) and (3.21) leads to the following relation between χ_{2D} and χ_{3D}:

$$\chi_{3D} \approx \frac{\chi_{2D}}{d}. \tag{3.22}$$

It follows that the relative permittivity of the homogenized slab is related to the metasurface susceptibility as

$$\epsilon_r = 1 + \chi_{3D} \approx 1 + \frac{\chi_{2D}}{d}, \tag{3.23}$$

where the division by the thickness d has the effect of "diluting the effect" of the susceptibility over the longitudinal extent of the metasurface. Note that χ_{2D} has the dimension of (m) since χ_{3D} is dimensionless. Similar relations can naturally be derived for the other scalar and tensorial susceptibilities.

3.3 Sheet Boundary Conditions

We will now discuss the three main techniques that have been developed to model the interactions of a metasurface with electromagnetic fields. As previously mentioned, these techniques are all based on the assumption that a metasurface may be reduced to an homogeneous zero-thickness sheet. We will thus assume that the conditions regarding the periodicity and thickness of the metasurface discussed in the previous two sections are satisfied.

3.3.1 Impedance Modeling

It is common practice in microwave electromagnetic theory to model structural discontinuities using impedance and admittance conditions. A well-known example is that of waveguide discontinuities, where the presence of irises or posts within the waveguide may be modeled using equivalent lumped elements [133].

The impedance boundary conditions (IBCs) based on the equivalence principle [54] are an extension of such methods. Figure 3.5 illustrates the concept of the IBCs.

Consider two regions separated by a surface S. Region 1, which is internal to S, supports the electric and magnetic fields $\mathbf{E}_1$ and $\mathbf{H}_1$. Region 2, which is external to S, supports the electric and magnetic fields $\mathbf{E}_2$ and $\mathbf{H}_2$. In general, S represents an electromagnetic discontinuity such that $\mathbf{E}_1 \neq \mathbf{E}_2$ and $\mathbf{H}_1 \neq \mathbf{H}_2$ at the boundary. The surface equivalence principle stipulates that the difference between these

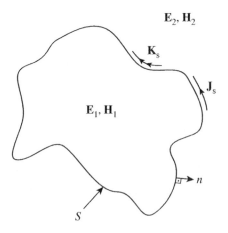

Figure 3.5 Electric and magnetic current densities, $\mathbf{J}_s$ and $\mathbf{K}_s$, on a surface S separating two regions supporting the electromagnetic fields $\mathbf{E}_1, \mathbf{H}_1$ and $\mathbf{E}_2, \mathbf{H}_2$, respectively.

fields may be related to equivalent surface current densities as

$$\mathbf{J}_s = \mathbf{n} \times (\mathbf{H}_2 - \mathbf{H}_1),$$ (3.24a)

$$\mathbf{K}_s = -\mathbf{n} \times (\mathbf{E}_2 - \mathbf{E}_1),$$ (3.24b)

where $\mathbf{J}_s$ and $\mathbf{K}_s$ are electric and magnetic surface currents on S. It is then possible to relate the average tangential electric (magnetic) field on S, $\mathbf{E}_{av}$ ($\mathbf{H}_{av}$), to the surface current $\mathbf{J}_s$ ($\mathbf{K}_s$) via [164]

$$Z_e \mathbf{J}_s = \mathbf{n} \times \mathbf{E}_{av},$$ (3.25a)

$$Y_m \mathbf{K}_s = \mathbf{n} \times \mathbf{H}_{av},$$ (3.25b)

where Z_e and Y_m are the electric impedance and the magnetic admittance, respectively. Since S is a zero-thickness sheet, the electric and magnetic fields on S may be approximated as the arithmetic averages of the fields at both sides of S, as discussed in Appendix 9.1, i.e.

$$\mathbf{E}_{av} = \frac{1}{2} (\mathbf{E}_1 + \mathbf{E}_2) \quad \text{and} \quad \mathbf{H}_{av} = \frac{1}{2} (\mathbf{H}_1 + \mathbf{H}_2).$$ (3.26)

Inserting (3.26) into (3.25) and using the result to eliminate the currents in (3.24), we get

$$\mathbf{n} \times \frac{(\mathbf{E}_1 + \mathbf{E}_2)}{2} = Z_e \mathbf{n} \times (\mathbf{H}_2 - \mathbf{H}_1),$$ (3.27a)

$$\mathbf{n} \times \frac{(\mathbf{H}_1 + \mathbf{H}_2)}{2} = -Y_m \mathbf{n} \times (\mathbf{E}_2 - \mathbf{E}_1).$$ (3.27b)

These expressions are one of the simplest forms of IBCs that apply to metasurfaces. They relate the discontinuities of the electric and magnetic fields produced by a metasurface to its corresponding effective impedance and admittance. These expressions are limited to isotropic structures and exclude coupling between electric and magnetic excitations, i.e. the electric (magnetic) surface current in (3.25) is only related to the electric (magnetic) field. However, we straightforwardly generalize the IBCs in (3.27) to the case of bianisotropic[2] metasurfaces as

$$\mathbf{n} \times \frac{(\mathbf{E}_2 + \mathbf{E}_1)}{2} = \overline{\overline{Z}}_{ee} \cdot \left[\mathbf{n} \times (\mathbf{H}_2 - \mathbf{H}_1) \right] + \overline{\overline{K}}_{me} \cdot \left[\mathbf{n} \times (\mathbf{E}_2 - \mathbf{E}_1) \right],$$ (3.28a)

$$\mathbf{n} \times \frac{(\mathbf{H}_2 + \mathbf{H}_1)}{2} = \overline{\overline{Y}}_{mm} \cdot \left[\mathbf{n} \times (\mathbf{E}_2 - \mathbf{E}_1) \right] + \overline{\overline{K}}_{em} \cdot \left[\mathbf{n} \times (\mathbf{H}_2 - \mathbf{H}_1) \right].$$ (3.28b)

where $\overline{\overline{Z}}_{ee}$, $\overline{\overline{Y}}_{mm}$, $\overline{\overline{K}}_{em}$, and $\overline{\overline{K}}_{me}$ are the electric surface impedance tensor, the magnetic surface admittance tensor, and the electric-to-magnetic and

---

2 The term bianisotropic refers to general tensorial constitutive material parameters, which include coupling between electric and magnetic responses. Refer to Section 2.1 for more information.

magnetic-to-electric coupling tensors, respectively [42]. For a metasurface lying in the xy-plane, the tensors in (3.28) are 2×2 matrices, with, for instance, $\overline{\overline{Z}}_{ee}$ reading

$$\overline{\overline{Z}}_{ee} = \begin{pmatrix} Z_{ee}^{xx} & Z_{ee}^{xy} \\ Z_{ee}^{yx} & Z_{ee}^{yy} \end{pmatrix}. \tag{3.29}$$

Thus, the modeling (3.28) takes into account only the tangential components of the electric and magnetic fields and the in-plane surface material parameters of the metasurface. This represents a restriction since the normal material parameters can also play a role in a metasurface [4]. We will get back to this restriction in the forthcoming sections and chapters. In Section 3.3.3, we will derive a more general and complete model, which notably takes into account both in-plane and out-of-plane material parameters.

3.3.2 Polarizability Modeling

The polarizability-based model first considers the scattering particles individually, and then combines them into a periodic subwavelength array to form the metasurface. Therefore, it directly relates the effect of the individual particles to the fields scattered by the metasurface. This is based on the assumption that the scattering particles are small enough compared to the illumination wavelength that they electromagnetically behave as equivalent electric and magnetic dipolar sources. In that case, the metasurface may then be seen as an equivalent array of coupled dipole moments. Summing the fields radiated by each of these dipolar sources yields the total fields scattered by the metasurface.

We here provide a brief description of this metasurface polarizability-based model. More details are available in [113]. In general, the equivalent dipole moments of an *isolated* scattering particle may be expressed as

$$\begin{pmatrix} \mathbf{p} \\ \mathbf{m} \end{pmatrix} = \begin{pmatrix} \overline{\overline{\alpha}}_{ee} & \overline{\overline{\alpha}}_{em} \\ \overline{\overline{\alpha}}_{me} & \overline{\overline{\alpha}}_{mm} \end{pmatrix} \cdot \begin{pmatrix} \mathbf{E}_{loc} \\ \mathbf{H}_{loc} \end{pmatrix}, \tag{3.30}$$

where $\mathbf{E}_{loc}$ and $\mathbf{H}_{loc}$ are the local electric and magnetic fields acting[3] on the scattering particle, and $\overline{\overline{\alpha}}_{ee}$, $\overline{\overline{\alpha}}_{mm}$, $\overline{\overline{\alpha}}_{em}$, and $\overline{\overline{\alpha}}_{me}$ are the electric, magnetic, magnetic-to-electric, and electric-to-magnetic polarizability tensors of the particle, respectively. Note that the system in (3.30) contains only the tangential components of the fields as well as the polarizabilities, which reduces the corresponding vectors and tensors to 2×1 vectors and 2×2 matrices, respectively.

---

3 The acting fields are defined as the total fields at the position of the scattering particle minus the fields scattered by the particle itself.

When an isolated scattering particle is inserted into a periodic array, the local fields in (3.30) correspond to a superposition of the incident field and the fields scattered by the adjacent particles, i.e. they include the coupling between the neighboring scattering particles. The local fields may then be expressed as [113, 164]

$$\mathbf{E}_{\text{loc}} = \mathbf{E}_{\text{inc}} + \bar{\bar{\beta}}_{\text{e}} \cdot \mathbf{p}, \tag{3.31a}$$

$$\mathbf{H}_{\text{loc}} = \mathbf{H}_{\text{inc}} + \bar{\bar{\beta}}_{\text{m}} \cdot \mathbf{m}, \tag{3.31b}$$

where $\bar{\bar{\beta}}_{\text{e}}$ and $\bar{\bar{\beta}}_{\text{m}}$ are the tensorial interaction constants

$$\bar{\bar{\beta}}_{\text{e}} = -\text{Re}\left[\frac{j\omega\eta_0}{4S}\left(1 - \frac{1}{jkR}\right)e^{-jkR}\right]\bar{\bar{I}}_{\text{t}} + j\left(\frac{\eta_0\epsilon_0\mu_0\omega^3}{6\pi} - \frac{\eta_0\omega}{2S}\right)\bar{\bar{I}}_{\text{t}}, \tag{3.32a}$$

$$\bar{\bar{\beta}}_{\text{m}} = \frac{\bar{\bar{\beta}}_{\text{e}}}{\eta_0^2}, \tag{3.32b}$$

where $S = L^2$ is the surface area of the unit cell, $\bar{\bar{I}}_{\text{t}}$ is the two-dimensional identity tensor, and $R = 0.6956L$ for a square periodic array [85]. It should be emphasized that the interaction constants in (3.32) are approximations that have been originally derived for normally propagating plane-wave illumination. Nevertheless, they may also be applied in the case of small angle oblique incidence. We may now express the dipole moments in (3.30) directly in terms of the incident fields by substituting (3.31) into (3.30), and solving for $\mathbf{p}$ and $\mathbf{m}$, which gives

$$\begin{pmatrix} \mathbf{p} \\ \mathbf{m} \end{pmatrix} = \begin{pmatrix} \bar{\bar{\alpha}}_{\text{ee}} & \bar{\bar{\alpha}}_{\text{em}} \\ \bar{\bar{\alpha}}_{\text{me}} & \bar{\bar{\alpha}}_{\text{mm}} \end{pmatrix} \cdot \begin{pmatrix} \mathbf{E}_{\text{inc}} \\ \mathbf{H}_{\text{inc}} \end{pmatrix}, \tag{3.33}$$

where the matrices $\bar{\bar{\alpha}}$ correspond to effective polarizabilities that represent the response from dipole moments of the scattering particles embedded in the metasurface array. In contrast to Eq. (3.30) where the dipole moments were expressed as functions of the acting fields, Eq. (3.33) provides a direct connection between the incident fields and the dipole moments via effective polarizability tensors. The latter are not provided here for brevity but may be found in [113].

A uniform metasurface may now be modeled by replacing each of its scattering particle by an equivalent dipole moment using (3.33), which results in a fictitious subwavelength array of dipole moments. It is then possible to relate these dipole moments to the fields scattered by the metasurface under a given illumination condition. The fields reflected and transmitted by the metasurface are found, following the derivation provided in Appendix 9.2, to be

$$\mathbf{E}_{\text{r}} = -\frac{j\omega}{2S}(\eta_0\mathbf{p} - \hat{z} \times \mathbf{m}), \tag{3.34a}$$

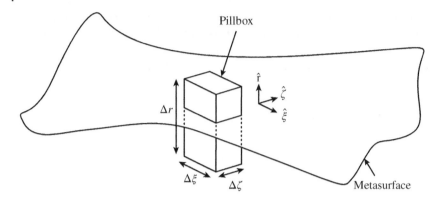

Figure 3.6 Curved metasurface with integration pillbox and corresponding system of coordinates.

$$\mathbf{E}_t = \mathbf{E}_{\text{inc}} - \frac{j\omega}{2S}(\eta_0 \mathbf{p} + \hat{\mathbf{z}} \times \mathbf{m}), \tag{3.34b}$$

respectively, with $\mathbf{p}$ and $\mathbf{m}$ given in (3.33) assuming a uniform metasurface.

Note that it is possible to directly connect the effective polarizabilities in (3.33) to the corresponding susceptibilities in (2.2). For that purpose, we provide mapping relations between polarizabilities and susceptibilities in Appendix 9.3.

3.3.3 Susceptibility Modeling

The susceptibility model that we shall discuss here is the most general and the most appropriate for metasurfaces. Compared to the previous two models, it naturally takes into account the presence of tangential and normal material parameters and is straightforwardly extendable to time-varying and nonlinear metasurfaces, as will be shown in forthcoming chapters. For this reason, most of the developments in this book will be based on this model.

The model consists in replacing a metasurface by an equivalent zero-thickness sheet of effective electric and magnetic surface polarization densities. These polarizations are then related to the effective surface susceptibilities of the metasurface. Originally developed by Idemen [64] in the seventies for the treatment of discontinuities in the electromagnetic field, the model was later applied to metasurfaces by Kuester [85]. Since then, it has been conventionally referred to as the Generalized Sheet Transition Conditions (GSTCs) and has been extensively applied to metasurfaces. For completeness, we shall derive it here directly from Maxwell equations, assuming a very general curved metasurface as the one shown in Figure 3.6. Note that alternative derivations are provided in [6, 64].

Let us start with the sourceless time-domain Maxwell equations,

$$\nabla \times \boldsymbol{\mathcal{E}} = -\frac{\partial}{\partial t}\boldsymbol{B}, \tag{3.35a}$$

$$\nabla \times \boldsymbol{\mathcal{H}} = \frac{\partial}{\partial t}\boldsymbol{D}, \tag{3.35b}$$

$$\nabla \cdot \boldsymbol{D} = 0, \tag{3.35c}$$

$$\nabla \cdot \boldsymbol{B} = 0. \tag{3.35d}$$

Alongside these equations, we also consider the constitutive relations,

$$\boldsymbol{D} = \epsilon_0 \boldsymbol{\mathcal{E}} + \boldsymbol{P}, \quad \text{or} \quad \boldsymbol{\mathcal{E}} = (\boldsymbol{D} - \boldsymbol{P})/\epsilon_0, \tag{3.36a}$$

$$\boldsymbol{B} = \mu_0(\boldsymbol{\mathcal{H}} + \boldsymbol{\mathcal{M}}), \quad \text{or} \quad \boldsymbol{\mathcal{H}} = \boldsymbol{B}/\mu_0 - \boldsymbol{\mathcal{M}}. \tag{3.36b}$$

Since the GSTCs model the effects of the metasurface scattering particles in terms of effective polarizations, we next use (3.36) to express $\boldsymbol{\mathcal{E}}$ and $\boldsymbol{D}$ in terms of $\boldsymbol{P}$ and $\boldsymbol{\mathcal{H}}$ and $\boldsymbol{B}$ in terms of $\boldsymbol{\mathcal{M}}$ in (3.35). This transforms (3.35) into

$$\nabla \times [(\boldsymbol{D} - \boldsymbol{P})/\epsilon_0)] = -\frac{\partial}{\partial t}\mu_0(\boldsymbol{\mathcal{H}} + \boldsymbol{\mathcal{M}}), \tag{3.37a}$$

$$\nabla \times (\boldsymbol{B}/\mu_0 - \boldsymbol{\mathcal{M}}) = \frac{\partial}{\partial t}(\epsilon_0 \boldsymbol{\mathcal{E}} + \boldsymbol{P}), \tag{3.37b}$$

$$\nabla \cdot \left(\epsilon_0 \boldsymbol{\mathcal{E}} + \boldsymbol{P}\right) = 0, \tag{3.37c}$$

$$\mu_0 \nabla \cdot (\boldsymbol{\mathcal{H}} + \boldsymbol{\mathcal{M}}) = 0. \tag{3.37d}$$

Next, we split the polarization densities into volume and surface components, i.e.

$$\boldsymbol{P} = \boldsymbol{P}_\text{v} + \boldsymbol{P}_\text{s}\delta(r), \tag{3.38a}$$

$$\boldsymbol{\mathcal{M}} = \boldsymbol{\mathcal{M}}_\text{v} + \boldsymbol{\mathcal{M}}_\text{s}\delta(r), \tag{3.38b}$$

where $\delta(r)$ is the Dirac delta distribution. Substituting (3.38) into (3.37), and placing the surface parts on the right-hand sides leads to

$$\nabla \times \left(\frac{\boldsymbol{D} - \boldsymbol{P}_\text{v}}{\epsilon_0}\right) = -\frac{\partial}{\partial t}\mu_0 \left[\boldsymbol{\mathcal{H}} + \boldsymbol{\mathcal{M}}_\text{v} + \boldsymbol{\mathcal{M}}_\text{s}\delta(r)\right] + \nabla \times \left[\frac{\boldsymbol{P}_\text{s}\delta(r)}{\epsilon_0}\right], \tag{3.39a}$$

$$\nabla \times \left(\frac{\boldsymbol{B}}{\mu_0} - \boldsymbol{\mathcal{M}}_\text{v}\right) = \frac{\partial}{\partial t}\left[\epsilon_0\boldsymbol{\mathcal{E}} + \boldsymbol{P}_\text{v} + \boldsymbol{P}_\text{s}\delta(r)\right] + \nabla \times [\boldsymbol{\mathcal{M}}_\text{s}\delta(r)], \tag{3.39b}$$

$$\nabla \cdot \left(\epsilon_0\boldsymbol{\mathcal{E}} + \boldsymbol{P}_\text{v}\right) = -\nabla \cdot \boldsymbol{P}_\text{s}\delta(r), \tag{3.39c}$$

$$\mu_0 \nabla \cdot \left(\boldsymbol{\mathcal{H}} + \boldsymbol{\mathcal{M}}_\text{v}\right) = -\mu_0 \nabla \cdot \boldsymbol{\mathcal{M}}_\text{s}\delta(r). \tag{3.39d}$$

We now make use of conventional pillbox integrations, and apply Stokes and Gauss theorems to obtain

$$\oint \left(\frac{\boldsymbol{D} - \boldsymbol{P}_\text{v}}{\epsilon_0}\right) \cdot d\mathbf{l} = -\mu_0 \frac{\partial}{\partial t}\iint (\boldsymbol{\mathcal{H}} + \boldsymbol{\mathcal{M}}_\text{v}) \cdot d\mathbf{S}$$
$$-\mu_0 \frac{\partial}{\partial t}\oiint \boldsymbol{\mathcal{M}}_\text{s}\delta(r) \cdot d\mathbf{S} + \oint \frac{\boldsymbol{P}_\text{s}}{\epsilon_0}\delta(r) \cdot d\mathbf{l}, \tag{3.40a}$$

$$\oint \left(\frac{\boldsymbol{B}}{\mu_0} - \boldsymbol{\mathcal{M}}_v \right) \cdot d\mathbf{l} = \epsilon_0 \frac{\partial}{\partial t} \iint \left(\boldsymbol{\mathcal{E}} + \frac{\boldsymbol{P}_v}{\epsilon_0} \right) \cdot d\mathbf{S}$$

$$+ \frac{\partial}{\partial t} \iint \boldsymbol{P}_s \delta(r) \cdot d\mathbf{S} + \oint \boldsymbol{\mathcal{M}}_s \delta(r) \cdot d\mathbf{l}, \qquad (3.40\text{b})$$

$$\oiint (\epsilon_0 \boldsymbol{\mathcal{E}} + \boldsymbol{P}_v) \cdot d\mathbf{S} = - \oiint \boldsymbol{P}_s \delta(r) \cdot d\mathbf{S}, \qquad (3.40\text{c})$$

$$\mu_0 \oiint (\boldsymbol{\mathcal{H}} + \boldsymbol{\mathcal{M}}_v) \cdot d\mathbf{S} = -\mu_0 \oiint \boldsymbol{\mathcal{M}}_s \delta(r) \cdot d\mathbf{S}, \qquad (3.40\text{d})$$

where $d\mathbf{S}$ and $d\mathbf{l}$ refer to the surfaces and contours of the pillbox in Figure 3.6, respectively. Noting that the integrands on the left-hand sides of (3.40) may be simplified using the constitutive relations (3.36), we substitute them (3.40) with

$$\frac{\boldsymbol{D}^\pm - \boldsymbol{P}_v^\pm}{\epsilon_0} = \boldsymbol{\mathcal{E}}^\pm, \quad \frac{\boldsymbol{B}^\pm}{\mu_0} - \boldsymbol{\mathcal{M}}_v^\pm = \boldsymbol{\mathcal{H}}^\pm, \quad \epsilon_0 \boldsymbol{\mathcal{E}}^\pm + \boldsymbol{P}_v^\pm = \boldsymbol{D}^\pm,$$

$$\mu_0 (\boldsymbol{\mathcal{H}}^\pm + \boldsymbol{\mathcal{M}}_v^\pm) = \boldsymbol{B}^\pm, \qquad (3.41)$$

where the minus signs refer to the $-\hat{\mathbf{r}}$ region, while the plus signs refer to the $+\hat{\mathbf{r}}$ region in Figure 3.6. Next, we integrate relations 3.40b and 3.40a over one of the ξr-plane of the pillbox, which yields the projections

$$(\mathcal{H}_\xi^+ - \mathcal{H}_\xi^-) \triangle \xi + (\mathcal{H}_r^{\text{left}} - \mathcal{H}_r^{\text{right}}) \triangle r = \epsilon_0 \frac{\partial}{\partial t} (\mathcal{E}_\zeta + P_{v,\zeta}/\epsilon_0) \triangle \xi \triangle r$$

$$+ \frac{\partial}{\partial t} P_{s,\zeta} \delta(r) \triangle \xi \triangle r - (M_{s,r}^{\text{right}} - M_{s,r}^{\text{left}}) \delta(r) \triangle r, \qquad (3.42\text{a})$$

$$(\mathcal{E}_\xi^+ - \mathcal{E}_\xi^-) \triangle \xi + (\mathcal{E}_r^{\text{left}} - \mathcal{E}_r^{\text{right}}) \triangle r = -\mu_0 \frac{\partial}{\partial t} (\mathcal{H}_\zeta + \mathcal{M}_{v,\zeta}) \triangle \xi \triangle r$$

$$-\mu_0 \frac{\partial}{\partial t} M_{s,\zeta} \delta(r) \triangle \xi \triangle r - (P_{s,r}^{\text{right}} - P_{s,r}^{\text{left}}) \delta(r) \triangle r/\epsilon_0, \qquad (3.42\text{b})$$

while their integration over one of the ζr-plane yields

$$(\mathcal{H}_\zeta^+ - \mathcal{H}_\zeta^-) \triangle \zeta - (\mathcal{H}_r^{\text{front}} - \mathcal{H}_r^{\text{back}}) \triangle r = -\epsilon_0 \frac{\partial}{\partial t} (\mathcal{E}_\xi + P_{v,\xi}/\epsilon_0) \triangle \zeta \triangle r$$

$$- \frac{\partial}{\partial t} P_{s,\xi} \delta(r) \triangle \zeta \triangle r - (M_{s,r}^{\text{front}} - M_{s,r}^{\text{back}}) \delta(r) \triangle r, \qquad (3.43\text{a})$$

$$(\mathcal{E}_\zeta^+ - \mathcal{E}_\zeta^-) \triangle \zeta - (\mathcal{E}_r^{\text{front}} - \mathcal{E}_r^{\text{back}}) \triangle r = \mu_0 \frac{\partial}{\partial t} (\mathcal{H}_\xi + \mathcal{M}_{v,\xi}) \triangle \zeta \triangle r$$

$$+\mu_0 \frac{\partial}{\partial t} M_{s,\xi} \delta(r) \triangle \zeta \triangle r - (P_{s,r}^{\text{front}} - P_{s,r}^{\text{back}}) \delta(r) \triangle r/\epsilon_0, \qquad (3.43\text{b})$$

where front and back refer to the $+\hat{\zeta}$ and $-\hat{\zeta}$ sides, and left and right refer to the $+\hat{\xi}$ and $-\hat{\xi}$ sides, respectively. The integration of relations 3.40c and 3.40d over the surfaces of the pillbox in Figure 3.6 yields

$$(D_r^+ - D_r^-) \triangle \xi \triangle \zeta + (D_\xi^{\text{right}} - D_\xi^{\text{left}}) \triangle r \triangle \zeta + (D_\zeta^{\text{front}} - D_\zeta^{\text{back}}{}_-) \triangle \xi \triangle r$$

$$= -\delta(r) \left[(P_{s,\xi}^{\text{right}} - P_{s,\xi}^{\text{left}}) \triangle \zeta \triangle r + (P_{s,\zeta}^{\text{front}} - P_{s,\zeta}^{\text{back}}) \triangle \xi \triangle r \right], \qquad (3.44\text{a})$$

$$(B_r^+ - B_r^-) \triangle \xi \triangle \zeta + (B_\xi^{\text{right}} - B_\xi^{\text{left}}) \triangle r \triangle \zeta + (B_\zeta^{\text{front}} - B_\zeta^{\text{back}}-) \triangle \xi \triangle r$$
$$= -\mu_0 \delta(r) \left[(\mathcal{M}_{s,\xi}^{\text{right}} - \mathcal{M}_{s,\xi}^{\text{left}}) \triangle \zeta \triangle r + (\mathcal{M}_{s,\zeta}^{\text{front}} - \mathcal{M}_{s,\zeta}^{\text{back}}) \triangle \xi \triangle r \right].$$

$$\text{(3.44b)}$$

To simplify expressions (3.42), (3.43), and (3.44), we take the limit $\triangle r \to 0$, replace $\delta(r) \triangle r \to 1$, and divide (3.42) by $\triangle \xi$, (3.43) by $\triangle \zeta$, and (3.44) by $\triangle \xi \triangle \zeta$, which yields

$$\triangle \mathcal{H}_\xi = \frac{\partial}{\partial t} \mathcal{P}_{s,\zeta} - \left(\mathcal{M}_{s,r}^{\text{right}} - \mathcal{M}_{s,r}^{\text{left}} \right) / \triangle \xi, \qquad (3.45a)$$

$$\triangle \mathcal{E}_\xi = -\mu_0 \frac{\partial}{\partial t} \mathcal{M}_{s,\zeta} - \left(\mathcal{P}_{s,r}^{\text{right}} - \mathcal{P}_{s,r}^{\text{left}} \right) / (\epsilon_0 \triangle \xi), \qquad (3.45b)$$

$$\triangle \mathcal{H}_\zeta = -\frac{\partial}{\partial t} \mathcal{P}_{s,\xi} - \left(\mathcal{M}_{s,r}^{\text{front}} - \mathcal{M}_{s,r}^{\text{back}} \right) / \triangle \zeta, \qquad (3.45c)$$

$$\triangle \mathcal{E}_\zeta = \mu_0 \frac{\partial}{\partial t} \mathcal{M}_{s,\xi} - \left(\mathcal{P}_{s,r}^{\text{front}} - \mathcal{P}_{s,r}^{\text{back}} \right) / (\epsilon_0 \triangle \zeta), \qquad (3.45d)$$

$$\triangle \mathcal{D}_r = - \left[\left(\mathcal{P}_{s,\xi}^{\text{right}} - \mathcal{P}_{s,\xi}^{\text{left}} \right) / \triangle \xi + \left(\mathcal{P}_{s,\zeta}^{\text{front}} - \mathcal{P}_{s,\zeta}^{\text{back}} \right) / \triangle \zeta \right], \qquad (3.45e)$$

$$\triangle \mathcal{B}_r = -\mu_0 \left[\left(\mathcal{M}_{s,\xi}^{\text{right}} - \mathcal{M}_{s,\xi}^{\text{left}} \right) / \triangle \xi + \left(\mathcal{M}_{s,\zeta}^{\text{front}} - \mathcal{M}_{s,\zeta}^{\text{back}} \right) / \triangle \zeta \right], \qquad (3.45f)$$

where $\triangle \mathcal{H}_\xi = \mathcal{H}_\xi^+ - \mathcal{H}_\xi^-$ and so on. Finally, taking the limits $\triangle \xi \to 0$ and $\triangle \zeta \to 0$, and combining 3.45a with 3.45c and 3.45b with 3.45d, yields the general expression of the time-domain GSTCs for a curved metasurface

$$\hat{\mathbf{r}} \times \triangle \mathcal{H} = \frac{\partial}{\partial t} \mathcal{P}_\parallel - \hat{\mathbf{r}} \times \nabla_\parallel \mathcal{M}_r, \qquad (3.46a)$$

$$\hat{\mathbf{r}} \times \triangle \mathcal{E} = -\mu_0 \frac{\partial}{\partial t} \mathcal{M}_\parallel - \hat{\mathbf{r}} \times \nabla_\parallel (\mathcal{P}_r / \epsilon_0), \qquad (3.46b)$$

$$\hat{\mathbf{r}} \cdot \triangle \mathcal{D} = -\nabla \cdot \mathcal{P}_\parallel, \qquad (3.46c)$$

$$\hat{\mathbf{r}} \cdot \triangle \mathcal{B} = -\mu_0 \nabla \cdot \mathcal{M}_\parallel, \qquad (3.46d)$$

where $\parallel$ refers to the tangential components ξ and ζ,[4] and where we have intentionally dropped the subscript "s," related to the surface polarization densities, for simplicity.

In a linear and/or time-invariant metasurface, the electric and magnetic polarization densities in the GSTCs may be expressed in terms of the susceptibilities and local fields using the relations (2.2). In the frequency domain, these fields are commonly approximated as the arithmetic average of the fields on both sides of

---

4 When the metasurface is flat, we consider that it is lying in the xy-plane at $z = 0$, which implies that $\hat{\mathbf{r}} = \hat{\mathbf{z}}$ and that $\parallel$ refers to $\hat{\mathbf{x}}$ and $\hat{\mathbf{y}}$.

the metasurface, as already done in Section 3.3.1 and discussed in Appendix 9.1. In this case, the relations (2.2) become

$$\mathbf{P} = \epsilon_0 \overline{\overline{\chi}}_{ee} \cdot \mathbf{E}_{av} + \frac{1}{c_0} \overline{\overline{\chi}}_{em} \cdot \mathbf{H}_{av}, \tag{3.47a}$$

$$\mathbf{M} = \overline{\overline{\chi}}_{mm} \cdot \mathbf{H}_{av} + \frac{1}{\eta_0} \overline{\overline{\chi}}_{me} \cdot \mathbf{E}_{av}, \tag{3.47b}$$

where the average fields are given in (3.26).

3.3.4 Comparisons Between the Models

We have presented three models for metasurfaces. These models are all electromagnetically valid and may all be used to synthesize and analyze metasurfaces. However, there are four important differences between them that we shall now discuss.

3.3.4.1 Microscopic and Macroscopic Perspectives

The IBCs and the GSTCs model metasurfaces in terms of macroscopic material parameters, while the polarizability model is based on microscopic quantities. Macroscopic parameters, such as the susceptibilities, are more related to the concept of a "medium" than the polarizabilities. Therefore, the susceptibilities are more intuitive since they can be straightforwardly related to well-known quantities, such as the permittivity and the refractive index. Similarly, impedances and admittances are familiar concepts to electrical engineers. On the other hand, polarizabilities are more ambiguous and their overall effects on the scattering behavior of a metasurface are harder to grasp since they are initially related to the response of single, isolated scatterers. To relate the scattering from a metasurface to the polarizabilities, one needs to consider the coupling tensors (3.32), which makes them less practical. However, the polarizabilities may be related to the geometry of the scattering particles more easily than macroscopic quantities. For instance, the dimensions of simple structures such as wires, loops, and omega-shaped particles have been successfully related to their corresponding polarizabilities [148, 164], which may be more difficult to achieve with macroscopic material parameters.

3.3.4.2 Material Tensor Dimensions and Normal Polarizations

The IBCs and the polarizability model use 2×2 material tensors, while the GSTCs use 3×3 tensors. The GSTCs include additional material parameters because they take into account the excitation of normal polarizations, as evidenced by the presence of the terms $\mathcal{P}_r$ and $\mathcal{M}_r$ in (3.46). It turns out that the presence of these normal polarizations may, in many cases, be ignored as their effect on the scattering of a metasurface is either negligible, non-existent, or may be modeled and thus replaced by purely tangential effective material parameters, as we shall discuss in Sec 6.4.1. In such a situation, the 3×3 material tensors used in the GSTCs do not present any advantage compared to the 2×2 tensors used by the other models. However, there are cases where these normal polarizations must be considered, thus making the GSTCs the only model able to be properly applied to these cases. We will present some of these cases in Chapters 4 and 6.

3.3.4.3 Uniform and Nonuniform Metasurfaces

The IBCs and the GSTCs relate the metasurface material parameters to the electromagnetic fields surrounding it in a very general fashion, since these fields may be defined arbitrarily. This is not the case of the polarizability model, since the expressions in (3.34) correspond to fields scattered by a uniform metasurface, as explained in Section 3.3.3. These expressions are not rigorous if the metasurface material parameters are spatially varying. In contrast, the IBCs and the GSTCs are valid for both uniform and nonuniform metasurfaces. Moreover, the coupling tensors (3.32) are approximate and rigorously apply only to normal plane-wave propagation [164]. This makes the IBCs and GSTCs more adequate for modeling operations such as oblique-wave or surface-wave propagation. Nevertheless, the polarizability model is perfectly applicable to normal-wave propagation, as well as nonuniform metasurfaces with slowly (with respect to the wavelength) spatially varying material parameters.

3.3.4.4 Extension to Time-Varying or Nonlinear Systems

The three models are easily extendable to time-varying systems by simply using their corresponding time-domain formulation. The inclusions of nonlinear responses are straightforward for the GSTCs since nonlinear polarizations and susceptibilities are commonly used in nonlinear optics. Similarly, nonlinear polarizabilities (or also referred to as hyper polarizabilities) have also been used for single scatterers, although they are less common than nonlinear susceptibilities. However, the concept of nonlinear impedances and admittances has, so far, not been applied to model nonlinear metasurfaces, which makes the IBCs less favorable for nonlinear systems.

From this comparison, we see that the GSTCs represent the most general model for metasurfaces, and it is thus the one that will be used in the forthcoming chapters.

4

Susceptibility Synthesis

The previous two chapters have introduced the main electromagnetic properties and modeling techniques pertaining to metasurfaces. We are now properly equipped to investigate how metasurfaces may be theoretically synthesized. This chapter provides, in Section 4.1, an extensive discussion on several aspects of the synthesis of linear and time-invariant (LTI) metasurfaces, while excluding the excitation of normal polarizations. Then, Section 4.1.6 extends that theory to the most general case of metasurfaces exhibiting both tangential and normal polarizations for completeness. Finally, Sections 4.2 and 4.3 discuss how the synthesis technique may be extended to the cases of time-varying and nonlinear metasurfaces, respectively.

4.1 Linear Time-Invariant Metasurfaces

4.1.1 Basic Assumptions

Synthesizing a linear time-invariant (LTI) metasurface is most conveniently achieved by expressing the GSTCs in the frequency domain. The frequency-domain GSTCs for a flat metasurface lying in the xy-plane at $z = 0$ are, from (3.46), given by

$$\hat{\mathbf{z}} \times \Delta \mathbf{H} = j\omega \mathbf{P}_{\parallel} - \hat{\mathbf{z}} \times \nabla_{\parallel} M_z, \tag{4.1a}$$

$$\hat{\mathbf{z}} \times \Delta \mathbf{E} = -j\omega \mu_0 \mathbf{M}_{\parallel} - \hat{\mathbf{z}} \times \nabla_{\parallel} (P_z/\epsilon_0), \tag{4.1b}$$

$$\hat{\mathbf{z}} \cdot \Delta \mathbf{D} = -\nabla \cdot \mathbf{P}_{\parallel}, \tag{4.1c}$$

$$\hat{\mathbf{z}} \cdot \Delta \mathbf{B} = -\mu_0 \nabla \cdot \mathbf{M}_{\parallel}, \tag{4.1d}$$

where the polarization densities $\mathbf{P}$ and $\mathbf{M}$ are given by (3.47) in terms of suscepti-bilities and average fields.

Electromagnetic Metasurfaces: Theory and Applications, First Edition.
Karim Achouri and Christophe Caloz.
© 2021 The Institute of Electrical and Electronics Engineers, Inc.
Published 2021 by John Wiley & Sons, Inc.

The metasurface synthesis procedure consists in solving the GSTC relations (4.1) to obtain the susceptibilities in (3.47), which are the unknowns of this inverse problem, so that the metasurface performs a desired electromagnetic transformation specified in terms of incident, reflected, and transmitted fields. According to the uniqueness theorem [140], Eqs. (4.1c) and (4.1d) are redundant relations in the absence of impressed sources since the transverse components of the fields are sufficient to completely describe the electromagnetic problem. Therefore, the metasurface synthesis operation is conventionally only performed with relations (4.1a) and (4.1b).

The GSTCs in (4.1) form a set of coupled nonhomogeneous partial differential equations with spatial derivatives applying to the normal components of the polarization densities in (4.1a) and (4.1b). Therefore, solving the inverse problem in the most general case, where all the susceptibility components are considered, is nontrivial and may require involved numerical analysis. Fortunately it is often appropriate, as will be discussed thereafter, to assume that the metasurface does not possess normal susceptibility components, i.e. $P_z = M_z = 0$. Thus, we shall next only consider cases where $P_z = M_z = 0$, which conveniently leads to closed-form solutions for the synthesized susceptibilities, while the more general case of nonzero normal susceptibilities will be addressed in Section 4.1.6.

Enforcing $P_z = M_z = 0$ may a priori seem an important restriction but, as we shall see, this does not have a major implication on the synthesis procedure, except in terms of reducing the number of the available degrees of freedom of the metasurface. Moreover, if a *uniform* metasurface is illuminated by a normally impinging plane wave, the presence of P_z and M_z does not affect the waves that it scatters. Indeed, in this case, both the fields and the metasurface susceptibilities are spatially uniform. Therefore, the spatial derivatives of P_z and M_z vanish in (4.1) and hence do not play a role in the response of the metasurface. Thus, in this scenario, the susceptibilities producing normal polarizations can be completely ignored; only the tangential components of the susceptibilities need to be considered.

There are at least three additional reasons that may be invoked for ignoring the presence of normal polarizations, which are the following: (i) Metasurfaces are often designed for waves propagating within the paraxial approximation, which implies that normal susceptibility components are much less excited than the tangential ones, since the fields are essentially transversely polarized. (ii) Because metasurfaces are electrically thin structures, excited normal polarizations have a negligible effect compared to the tangential ones. (iii) Since, from the Huygens principle, electromagnetic fields can always be expressed in terms of tangential components, a metasurface possessing both normal and tangential susceptibility components may be transformed into an equivalent metasurface with only tangential susceptibility components [10]. This point is illustrated in Figure 4.1a and b, where the scattering response of a metasurface with scatterers exhibiting

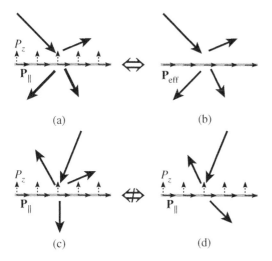

Figure 4.1 Illustrations of the application of the Huygens principle and its limitation. (a) Metasurface with complex orientations of scattering particles and resulting vertical and horizontal polarizations. (b) Metasurface equivalent to (a) but with purely tangential dipoles. (c) The metasurface in (a) is now excited at other incidence angle and produces different scattered fields than in (a). (d) The purely tangential metasurface (b) is not the equivalent of the one in (c) anymore since different normal polarizations are induced [5]. In these illustrations, only electric polarizations are depicted for convenience but magnetic ones may also naturally exist.

both normal and tangential polarizations is identical to that of an equivalent metasurface with scatterers producing only tangential polarizations. This may be understood by realizing that the effects of $\mathbf{P}_{\parallel}$ ($\mathbf{M}_{\parallel}$) and the gradient of M_z (P_z) in Eqs. (4.1a) and (4.1b) may be combined into a single effective parameter $\mathbf{P}_{\text{eff}}$ ($\mathbf{M}_{\text{eff}}$).

While the three points mentioned above are valid in most cases, there are situations where they do not apply and are thus invalid. For instance, a metasurface may be designed for waves propagating at grazing angles, in which case the normal components of the fields dominate over the tangential ones, implying that normal susceptibility component may be strongly excited. It is also possible that even purely planar scattering particles may exhibit normal polarizations, which is typically the case of in-plane conductive loops such as split-ring resonators. Finally, the application of the Huygens principle illustrated in Figure 4.1a and b is only valid if the direction of propagation of the incident wave is identical for both surfaces. Indeed, since the gradient of M_z (P_z) in Eqs. (4.1a) and (4.1b) is dependent on the direction of wave propagation, it is generally impossible to find an effective parameter $\mathbf{P}_{\text{eff}}$ ($\mathbf{M}_{\text{eff}}$) that would be the same irrespective of the incidence angle. This is illustrated in Figure 4.1c and d, where the metasurfaces are the same as those in Figure 4.1a and b, respectively, but where the changed angle of incidence leads to different scattering responses for the two structures.

In summary, the presence of normal polarizations may play a non-negligible role and should thus be accounted for *in the design of metasurfaces meant to be used at different angles of incidence.* For metasurfaces designed for a specific incidence angle, then the presence and the effects of normal polarizations may be modeled by effective tangential polarizations, according to the Huygens principle, and can thus be ignored [5].

Let us now simplify the GSTCs so as to obtain the final form of the synthesis relations in the absence of normal polarizations. Substituting (3.47) into (4.1a) and (4.1b), and dropping the spatial derivatives, leads to

$$\hat{z} \times \Delta \mathbf{H} = j\omega\epsilon_0 \overline{\overline{\chi}}_{ee} \cdot \mathbf{E}_{av} + jk_0 \overline{\overline{\chi}}_{em} \cdot \mathbf{H}_{av}, \tag{4.2a}$$

$$\hat{z} \times \Delta \mathbf{E} = -j\omega\mu_0 \overline{\overline{\chi}}_{mm} \cdot \mathbf{H}_{av} - jk_0 \overline{\overline{\chi}}_{me} \cdot \mathbf{E}_{av}, \tag{4.2b}$$

where k_0 is the free-space wavenumber and where the susceptibility tensors only contain tangential susceptibility components. This system may be expressed in the following matrix form to simplify the synthesis procedure:

$$\begin{pmatrix} \Delta H_y \\ \Delta H_x \\ \Delta E_y \\ \Delta E_x \end{pmatrix} = \begin{pmatrix} \hat{\chi}_{ee}^{xx} & \hat{\chi}_{ee}^{xy} & \hat{\chi}_{em}^{xx} & \hat{\chi}_{em}^{xy} \\ \hat{\chi}_{ee}^{yx} & \hat{\chi}_{ee}^{yy} & \hat{\chi}_{em}^{yx} & \hat{\chi}_{em}^{yy} \\ \hat{\chi}_{me}^{xx} & \hat{\chi}_{me}^{xy} & \hat{\chi}_{mm}^{xx} & \hat{\chi}_{mm}^{xy} \\ \hat{\chi}_{me}^{yx} & \hat{\chi}_{me}^{yy} & \hat{\chi}_{mm}^{yx} & \hat{\chi}_{mm}^{yy} \end{pmatrix} \cdot \begin{pmatrix} E_{x,av} \\ E_{y,av} \\ H_{x,av} \\ H_{y,av} \end{pmatrix}, \tag{4.3}$$

where the hat symbol indicates that the susceptibilities in (4.2) have been normalized. The relationships between the susceptibilities in (4.2) and those in (4.3) are

$$\overline{\overline{\chi}}_{ee} = -j\omega\epsilon_0 \overline{\overline{N}} \cdot \overline{\overline{\hat{\chi}}}_{ee} \quad \leftrightarrow \quad \overline{\overline{\hat{\chi}}}_{ee} = \frac{j}{\omega\epsilon_0} \overline{\overline{N}} \cdot \overline{\overline{\chi}}_{ee}, \tag{4.4a}$$

$$\overline{\overline{\chi}}_{mm} = j\omega\mu_0 \overline{\overline{N}} \cdot \overline{\overline{\hat{\chi}}}_{mm} \quad \leftrightarrow \quad \overline{\overline{\hat{\chi}}}_{mm} = -\frac{j}{\omega\mu_0} \overline{\overline{N}} \cdot \overline{\overline{\chi}}_{mm}, \tag{4.4b}$$

$$\overline{\overline{\chi}}_{em} = -jk_0 \overline{\overline{N}} \cdot \overline{\overline{\hat{\chi}}}_{em} \quad \leftrightarrow \quad \overline{\overline{\hat{\chi}}}_{em} = \frac{j}{k_0} \overline{\overline{N}} \cdot \overline{\overline{\chi}}_{em}, \tag{4.4c}$$

$$\overline{\overline{\chi}}_{me} = jk_0 \overline{\overline{N}} \cdot \overline{\overline{\hat{\chi}}}_{me} \quad \leftrightarrow \quad \overline{\overline{\hat{\chi}}}_{me} = -\frac{j}{k_0} \overline{\overline{N}} \cdot \overline{\overline{\chi}}_{me}, \tag{4.4d}$$

where

$$\overline{\overline{N}} = \begin{pmatrix} 1 & 0 \\ 0 & -1 \end{pmatrix}. \tag{4.5}$$

Note that in (4.4), the susceptibility tensors are 2×2 matrices containing only the tangential components shown in (4.3).

A quick inspection of the system (4.3) reveals that it contains 16 unknown susceptibilities for only four equations. This means that it is heavily underdetermined and can therefore not be solved directly to obtain the susceptibilities in terms of the specified fields. This indeterminacy prompts for two distinct resolution strategies. The first is to reduce to four the number of independent unknowns so as to obtain a full-rank system. The fact that different quadruplets of susceptibility components may be selected indicates that different combinations of susceptibilities produce the exact same scattered fields. The second is to increase the number of specified transformations. This means that the metasurface has the capability to independently transform several sets of incident, reflected, and transmitted waves. There are therefore three main methods to solve the inverse synthesis problem: (i) reducing the number of independent unknowns, (ii) increasing the number of transformations, or (iii) a combination of (i) and (ii).

To reduce the number of unknowns, one may, for instance, enforce susceptibility component interdependence conditions. Such conditions include the reciprocity conditions in (2.52), and the gainlessness and losslessness conditions in (2.77), which reduce the number of independent variables due to the interdependence of several susceptibility components. For instance, the reciprocity conditions reduce the number of independent tangential susceptibilities from 16 to 10. However, this strategy may not be the most appropriate since the conditions that are imposed may not be compatible with the design or cost constraints.

The most general approach to solve the system (4.3) is to match the number of unknown susceptibilities to the number of specified transformations. In many cases, only one transformation is required and thus only four susceptibilities are used to synthesize the metasurface.[1] With 4 susceptibilities out of 16, the number of possible combinations of susceptibilities is very large. However, most of these combinations lead to nonphysical or unpractical designs. This point will be illustrated in Section 4.1.7.1, where we will show how polarization rotation can be achieved by considering different combinations of susceptibilities. From a general perspective, it is clear that the choice of susceptibilities depends on the requirements of the specified problem. Note that these considerations are naturally extended to the cases where more than one transformation are desired.

In the forthcoming discussions, we will only present a limited number of combinations of susceptibilities, for the sake of conciseness, but without loss of generality. We will start by considering the synthesis of birefringent metasurfaces,

---

1 Note that a single transformation between specified incident, reflected, and transmitted waves requires four susceptibilities only in the general case where these waves exhibit both x and y polarization states. However, if only one of these two polarization states is considered, then the system (4.3) reduces to two equations. In this scenario, only two susceptibilities instead of four are required to synthesize the metasurface.

which is one of the most common types of metasurfaces. Then, we will present an example of multiple transformations.

4.1.2 Birefringent Metasurfaces

One of the simplest and most common metasurfaces is the homoanisotropic metasurface, which possesses only diagonal nonzero susceptibility tensors. This is a birefringent structure [143], for which the system (4.3) reduces to

$$
\begin{pmatrix} \Delta H_y \\ \Delta H_x \\ \Delta E_y \\ \Delta E_x \end{pmatrix} = \begin{pmatrix} \hat{\chi}_{ee}^{xx} & 0 & 0 & 0 \\ 0 & \hat{\chi}_{ee}^{yy} & 0 & 0 \\ 0 & 0 & \hat{\chi}_{mm}^{xx} & 0 \\ 0 & 0 & 0 & \hat{\chi}_{mm}^{yy} \end{pmatrix} \cdot \begin{pmatrix} E_{x,\mathrm{av}} \\ E_{y,\mathrm{av}} \\ H_{x,\mathrm{av}} \\ H_{y,\mathrm{av}} \end{pmatrix}.
\tag{4.6}
$$

This system may be straightforwardly solved and yields, using (4.4), the following closed-form relations for the four susceptibilities:

$$
\chi_{ee}^{xx} = \frac{-\Delta H_y}{j\omega\epsilon_0 E_{x,\mathrm{av}}},
\tag{4.7a}
$$

$$
\chi_{ee}^{yy} = \frac{\Delta H_x}{j\omega\epsilon_0 E_{y,\mathrm{av}}},
\tag{4.7b}
$$

$$
\chi_{mm}^{xx} = \frac{\Delta E_y}{j\omega\mu_0 H_{x,\mathrm{av}}},
\tag{4.7c}
$$

$$
\chi_{mm}^{yy} = \frac{-\Delta E_x}{j\omega\mu_0 H_{y,\mathrm{av}}},
\tag{4.7d}
$$

where $\Delta H_y = H_{y,t} - (H_{y,i} + H_{y,r})$, $E_{x,\mathrm{av}} = (E_{x,t} + E_{x,i} + E_{x,r})/2$, and so on, with i, r, and t indicating the fields of the incident, reflected, and transmitted waves, respectively. A metasurface with the susceptibilities in (4.7) will exactly produce the specified reflected and transmitted transverse components of the fields when the metasurface is illuminated by the specified incident field. Since the longitudinal fields are completely determined from the transverse components, according to the uniqueness theorem, the complete specified electromagnetic fields are exactly generated by the metasurface.

Due to the orthogonality between x- and y-polarized waves, the susceptibilities in (4.7) separate in two subsets, corresponding to Eqs. (4.7a) and (4.7d), and Eqs. (4.7b) and (4.7c). These two sets of susceptibilities independently and simultaneously transform x- and y-polarized waves, so that each subset performs the simplest possible single transformation. If the two electric and the two magnetic susceptibilities in (4.7) are equal to each other ($\chi_{ee}^{xx} = \chi_{ee}^{yy}$ and $\chi_{mm}^{xx} = \chi_{mm}^{yy}$), then the metasurface is homoisotropic and hence performs the same operation for x- and y-polarized waves. If this is not the case, then the metasurface is homoanisotropic and hence birefringent, performing the simplest possible double transformation. We will see in the next section a more general case of multiple transformation that is not based on an orthogonal separation of x- and y-polarized waves as is the case here.

The system in (4.6) represents the simplest way to synthesize metasurfaces performing single (or double) transformation, but it is obviously not the only possible way. One may for instance imagine a homoanisotropic metasurface with nonzero *off-diagonal* components, which would be solved in the exact same fashion while performing a different kind of electromagnetic transformation involving gyrotropy. An example of such a gyrotropic metasurface will be discussed in Section 4.1.7.1.

4.1.3 Multiple-Transformation Metasurfaces

We have just seen how a metasurface can be synthesized to perform a single transformation, if the structure is homoisotropic, or a birefringent double transformation, if the structure is homoanisotropic. However, the general system of equations (4.3) points to the possibility of multiple transformations given its large number of degrees of freedom, specifically its 16 susceptibility components. In fact, the system (4.3) can be solved for several transformations, which, for instance, includes incident waves impinging on the metasurface from one of its side only or both. Up to three additional wave transformations can be added to obtain a full-rank system. Accordingly, we have that

$$
\begin{pmatrix}
\Delta H_{y1} & \Delta H_{y2} & \Delta H_{y3} & \Delta H_{y4} \\
\Delta H_{x1} & \Delta H_{x2} & \Delta H_{x3} & \Delta H_{x4} \\
\Delta E_{y1} & \Delta E_{y2} & \Delta E_{y3} & \Delta E_{y4} \\
\Delta E_{x1} & \Delta E_{x2} & \Delta E_{x3} & \Delta E_{x4}
\end{pmatrix} =
$$

$$
\begin{pmatrix}
\hat{\chi}_{ee}^{xx} & \hat{\chi}_{ee}^{xy} & \hat{\chi}_{em}^{xx} & \hat{\chi}_{em}^{xy} \\
\hat{\chi}_{ee}^{yx} & \hat{\chi}_{ee}^{yy} & \hat{\chi}_{em}^{yx} & \hat{\chi}_{em}^{yy} \\
\hat{\chi}_{me}^{xx} & \hat{\chi}_{me}^{xy} & \hat{\chi}_{mm}^{xx} & \hat{\chi}_{mm}^{xy} \\
\hat{\chi}_{me}^{yx} & \hat{\chi}_{me}^{yy} & \hat{\chi}_{mm}^{yx} & \hat{\chi}_{mm}^{yy}
\end{pmatrix} \cdot
\begin{pmatrix}
E_{x1,av} & E_{x2,av} & E_{x3,av} & E_{x4,av} \\
E_{y1,av} & E_{y2,av} & E_{y3,av} & E_{y4,av} \\
H_{x1,av} & H_{x2,av} & H_{x3,av} & H_{x4,av} \\
H_{y1,av} & H_{y2,av} & H_{y3,av} & H_{y4,av}
\end{pmatrix},
$$

$$(4.8)$$

where the subscripts 1, 2, 3, and 4 label the four distinct and *independent* sets of waves. As in previous expressions, the susceptibilities can be obtained by matrix inversion, conjointly with (4.4).

For the sake of illustration, we now consider a specific example of a double transformation by a homoanisotropic metasurface with full electric and magnetic tangential susceptibility tensors. The corresponding system is, from (4.8), given by

$$
\begin{pmatrix}
\Delta H_{y1} & \Delta H_{y2} \\
\Delta H_{x1} & \Delta H_{x2} \\
\Delta E_{y1} & \Delta E_{y2} \\
\Delta E_{x1} & \Delta E_{x2}
\end{pmatrix}
=
\begin{pmatrix}
\hat{\chi}_{ee}^{xx} & \hat{\chi}_{ee}^{xy} & 0 & 0 \\
\hat{\chi}_{ee}^{yx} & \hat{\chi}_{ee}^{yy} & 0 & 0 \\
0 & 0 & \hat{\chi}_{mm}^{xx} & \hat{\chi}_{mm}^{xy} \\
0 & 0 & \hat{\chi}_{mm}^{yx} & \hat{\chi}_{mm}^{yy}
\end{pmatrix}
\cdot
\begin{pmatrix}
E_{x1,av} & E_{x2,av} \\
E_{y1,av} & E_{y2,av} \\
H_{x1,av} & H_{x2,av} \\
H_{y1,av} & H_{y2,av}
\end{pmatrix}.
\tag{4.9}
$$

We assume that the two transformations that we are considering possess fields with both x and y polarizations. The solution of (4.9) is readily found by matrix inversion. Using (4.4), the resulting susceptibilities are

$$
\chi_{ee}^{xx} = \frac{j}{\epsilon_0 \omega} \frac{(E_{y1,av}\Delta H_{y2} - E_{y2,av}\Delta H_{y1})}{(E_{x2,av}E_{y1,av} - E_{x1,av}E_{y2,av})},
\tag{4.10a}
$$

$$
\chi_{ee}^{xy} = \frac{j}{\epsilon_0 \omega} \frac{(E_{x2,av}\Delta H_{y1} - E_{x1,av}\Delta H_{y2})}{(E_{x2,av}E_{y1,av} - E_{x1,av}E_{y2,av})},
\tag{4.10b}
$$

$$
\chi_{ee}^{yx} = \frac{j}{\epsilon_0 \omega} \frac{(E_{y2,av}\Delta H_{x1} - E_{y1,av}\Delta H_{x2})}{(E_{x2,av}E_{y1,av} - E_{x1,av}E_{y2,av})},
\tag{4.10c}
$$

$$
\chi_{ee}^{yy} = \frac{j}{\epsilon_0 \omega} \frac{(E_{x1,av}\Delta H_{x2} - E_{x2,av}\Delta H_{x1})}{(E_{x2,av}E_{y1,av} - E_{x1,av}E_{y2,av})},
\tag{4.10d}
$$

$$
\chi_{mm}^{xx} = \frac{j}{\mu_0 \omega} \frac{(H_{y2,av}\Delta E_{y1} - H_{y1,av}\Delta E_{y2})}{(H_{x2,av}H_{y1,av} - H_{x1,av}H_{y2,av})},
\tag{4.10e}
$$

$$
\chi_{mm}^{xy} = \frac{j}{\mu_0 \omega} \frac{(H_{x1,av}\Delta E_{y2} - H_{x2,av}\Delta E_{y1})}{(H_{x2,av}H_{y1,av} - H_{x1,av}H_{y2,av})},
\tag{4.10f}
$$

$$
\chi_{mm}^{yx} = \frac{j}{\mu_0 \omega} \frac{(H_{y1,av}\Delta E_{x2} - H_{y2,av}\Delta E_{x1})}{(H_{x2,av}H_{y1,av} - H_{x1,av}H_{y2,av})},
\tag{4.10g}
$$

$$
\chi_{mm}^{yy} = \frac{j}{\mu_0 \omega} \frac{(H_{x2,av}\Delta E_{x1} - H_{x1,av}\Delta E_{x2})}{(H_{x2,av}H_{y1,av} - H_{x1,av}H_{y2,av})},
\tag{4.10h}
$$

where the subscripts 1 and 2 label the first and the second wave set transformation, respectively. Applying the conditions (2.52) and (2.77) to (4.10) indicates that the metasurface may generally be both active/lossy and nonreciprocal. The same argument applies to the more general case of the fully bianisotropic metasurface, described by the susceptibilities in (4.8), which may, depending on the choice of transformations, be nonreciprocal and active/lossy. Note that the choice of using the susceptibility tensors $\overline{\overline{\chi}}_{ee}$ and $\overline{\overline{\chi}}_{mm}$ in (4.9) was arbitrary and other sets of susceptibilities, for instance including bianisotropic components, may be better suited to perform some transformations. Note that an application example of the relations (4.10) will be presented in Section 4.1.7.2.

4.1.4 Relations Between Susceptibilities and Scattering Parameters

So far, we have been interested only in finding the susceptibilities in terms of specified fields. We shall now investigate how these synthesized susceptibilities may be related to the shape of the scattering particles that will constitute the metasurfaces to be realized. This is an essential step toward the practical realization of a metasurface. While a general discussion on the realization of metasurface scattering particles will be presented in Chapter 6, we will here only present the mathematical expressions that relate the susceptibilities to the scattering response of scattering particles.

The conventional method to relate the scattering particle shape to equivalent susceptibilities (or material parameters) is based on homogenization techniques. In the case of metamaterials, these techniques may be used to relate homogenized material parameters to the scattering parameters of the scatterers. In general, a single isolated scatterer is not sufficient to describe an homogenized medium. Instead, one should consider a periodic array of scatterers that takes into account the coupling interactions between adjacent scatterers to obtain an accurate "medium" description. The susceptibilities, which describe the macroscopic responses of a medium, are thus naturally suited to describe the homogenized material parameters of metasurfaces. It follows that the equivalent susceptibilities of a scattering particle may be related to their scattering parameters, which are typically obtained by periodic full-wave simulations [14, 15, 125, 129]. Because the periodic array of scatterers is uniform with subwavelength periodicity, the scattered fields obey Snell's law. Specifically, if the incident wave impinges normally onto the metasurface, then the reflected and transmitted waves also propagate normally, as discussed in Section 3.1.3. When the periodic array of scattering particles is excited normally, one can rigorously obtain the 16 *tangential* susceptibility components in (4.8). However, this does not provide any information on the normal susceptibility components of the scattering particles. This is because the excitation of normal polarizations, in the case of normally propagating waves, does not induce any discontinuity in the fields, if the metasurface is uniform. Nevertheless, this method allows one to match the tangential susceptibilities of the scattering particle to the susceptibilities found from the metasurface synthesis procedure and this precisely yields the ideal tangential susceptibility components. It is clear that the scattering particles may, in addition to their tangential susceptibilities, possess nonzero normal susceptibility components. In this case, the scattering response of the metasurface will differ from the expected ideal behavior prescribed in the synthesis, when illuminated with obliquely propagating waves. Thus, the homogenization technique only serves as an initial guess to describe the scattering behavior of the metasurface. Note that is possible to obtain the 36 susceptibility components (4 susceptibility tensors, each with 9 components) of a scattering particle, but that would require solving the 4 GSTC relations for 9 independent sets of incident, reflected, and transmitted waves, which is particularly tedious and hence generally avoided.

We shall now derive the expressions relating the tangential susceptibilities to the scattering parameters in the general case of a fully bianisotropic uniform metasurface surrounded by different media and excited by normally incident plane waves. Let us first write the system (4.8) in the compact form

$$\overline{\overline{\Delta}} = \overline{\overline{\chi}} \cdot \overline{\overline{A}}_v, \tag{4.11}$$

where the matrices $\overline{\overline{\Delta}}$, $\overline{\overline{\chi}}$ and $\overline{\overline{A}}_v$ correspond to the field differences, the normalized susceptibilities, and the field averages, respectively. In order to determine the 16 tangential susceptibility components in (4.8), we define four transformations and specify their fields on both sides of the metasurface. Let us consider that the metasurface is illuminated with an x-polarized normally incident plane wave propagating in the positive z-direction. The corresponding incident, reflected, and transmitted electromagnetic fields read

$$\mathbf{E}_i = \hat{\mathbf{x}}, \quad \mathbf{E}_r = S_{11}^{xx}\hat{\mathbf{x}} + S_{11}^{yx}\hat{\mathbf{y}}, \quad \mathbf{E}_t = S_{21}^{xx}\hat{\mathbf{x}} + S_{21}^{yx}\hat{\mathbf{y}}, \tag{4.12a}$$

$$\mathbf{H}_i = \frac{1}{\eta_1}\hat{\mathbf{y}}, \quad \mathbf{H}_r = \frac{1}{\eta_1}(S_{11}^{yx}\hat{\mathbf{x}} - S_{11}^{xx}\hat{\mathbf{y}}), \quad \mathbf{H}_t = \frac{1}{\eta_2}(-S_{21}^{yx}\hat{\mathbf{x}} + S_{21}^{xx}\hat{\mathbf{y}}), \tag{4.12b}$$

where the terms S_{ab}^{uv}, with $a,b = \{1,2\}$ and $u,v = \{x,y\}$, are the scattering parameters with ports 1 and 2 corresponding to the left $(z = 0^-)$ and right $(z = 0^+)$ sides of the metasurface, respectively. The medium on the left of the metasurface has the intrinsic impedance η_1, while the medium on the right has the intrinsic impedance η_2. In addition to (4.12), three other cases have to be considered, namely y-polarized excitation incident from the left (port 1), and x- and y-polarized excitations incident from the right (port 2). Inserting these fields into (4.8) leads to the matrix

$$\overline{\overline{\Delta}} = \begin{pmatrix} -\overline{\overline{N}}/\eta_1 + \overline{\overline{N}} \cdot \overline{\overline{S}}_{11}/\eta_1 + \overline{\overline{N}} \cdot \overline{\overline{S}}_{21}/\eta_2 & -\overline{\overline{N}}/\eta_2 + \overline{\overline{N}} \cdot \overline{\overline{S}}_{12}/\eta_1 + \overline{\overline{N}} \cdot \overline{\overline{S}}_{22}/\eta_2 \\ -\overline{\overline{J}} \cdot \overline{\overline{N}} - \overline{\overline{J}} \cdot \overline{\overline{N}} \cdot \overline{\overline{S}}_{11} + \overline{\overline{J}} \cdot \overline{\overline{N}} \cdot \overline{\overline{S}}_{21} & \overline{\overline{J}} \cdot \overline{\overline{N}} - \overline{\overline{J}} \cdot \overline{\overline{N}} \cdot \overline{\overline{S}}_{12} + \overline{\overline{J}} \cdot \overline{\overline{N}} \cdot \overline{\overline{S}}_{22} \end{pmatrix}, \tag{4.13}$$

with

$$\overline{\overline{A}}_v = \frac{1}{2}\begin{pmatrix} \overline{\overline{I}} + \overline{\overline{S}}_{11} + \overline{\overline{S}}_{21} & \overline{\overline{I}} + \overline{\overline{S}}_{12} + \overline{\overline{S}}_{22} \\ \overline{\overline{J}}/\eta_1 - \overline{\overline{J}} \cdot \overline{\overline{S}}_{11}/\eta_1 + \overline{\overline{J}} \cdot \overline{\overline{S}}_{21}/\eta_2 & -\overline{\overline{J}}/\eta_2 - \overline{\overline{J}} \cdot \overline{\overline{S}}_{12}/\eta_1 + \overline{\overline{J}} \cdot \overline{\overline{S}}_{22}/\eta_2 \end{pmatrix}, \tag{4.14}$$

where

$$\overline{\overline{S}}_{ab} = \begin{pmatrix} S_{ab}^{xx} & S_{ab}^{xy} \\ S_{ab}^{yx} & S_{ab}^{yy} \end{pmatrix}, \quad \overline{\overline{I}} = \begin{pmatrix} 1 & 0 \\ 0 & 1 \end{pmatrix}, \quad \overline{\overline{J}} = \begin{pmatrix} 0 & -1 \\ 1 & 0 \end{pmatrix}, \quad \overline{\overline{N}} = \begin{pmatrix} 1 & 0 \\ 0 & -1 \end{pmatrix}. \tag{4.15}$$

The general procedure to obtain the susceptibilities of a given scattering particle will be detailed in Section 6.1 but may be summarized as follows: Firstly, simulate the scattering particle with periodic boundary conditions and for normal excitation. Secondly, use the resulting scattering parameters to define the matrices in (4.13) and (4.14). Finally, the corresponding susceptibilities are obtained by matrix inversion of (4.11).

By reversing (4.11) and solving for the scattering parameters, it is alternatively possible to express the scattering parameters (for normal-wave propagation) of a uniform metasurface with known susceptibilities. This leads to the matrix equation

$$\bar{\bar{S}} = \bar{\bar{M}}_1^{-1} \cdot \bar{\bar{M}}_2, \tag{4.16}$$

where the scattering parameters matrix, $\bar{\bar{S}}$, is defined as

$$\bar{\bar{S}} = \begin{pmatrix} \bar{\bar{S}}_{11} & \bar{\bar{S}}_{12} \\ \bar{\bar{S}}_{21} & \bar{\bar{S}}_{22} \end{pmatrix}, \tag{4.17}$$

and the matrices $\bar{\bar{M}}_1$ and $\bar{\bar{M}}_2$ are obtained from (4.11), (4.13), and (4.14) by expressing the scattering parameters in terms of the normalized susceptibility tensors. The resulting matrices are

$$\bar{\bar{M}}_1 = \begin{pmatrix} \bar{\bar{N}}/\eta_1 - \bar{\bar{\chi}}_{ee}/2 + \bar{\bar{\chi}}_{em} \cdot \bar{\bar{J}}/(2\eta_1) & \bar{\bar{N}}/\eta_2 - \bar{\bar{\chi}}_{ee}/2 - \bar{\bar{\chi}}_{em} \cdot \bar{\bar{J}}/(2\eta_2) \\ -\bar{\bar{J}} \cdot \bar{\bar{N}} - \bar{\bar{\chi}}_{me}/2 + \bar{\bar{\chi}}_{mm} \cdot \bar{\bar{J}}/(2\eta_1) & \bar{\bar{J}} \cdot \bar{\bar{N}} - \bar{\bar{\chi}}_{me}/2 - \bar{\bar{\chi}}_{mm} \cdot \bar{\bar{J}}/(2\eta_2) \end{pmatrix}, \tag{4.18}$$

and

$$\bar{\bar{M}}_2 = \begin{pmatrix} \bar{\bar{\chi}}_{ee}/2 + \bar{\bar{N}}/\eta_1 + \bar{\bar{\chi}}_{em} \cdot \bar{\bar{J}}/(2\eta_1) & \bar{\bar{\chi}}_{ee}/2 + \bar{\bar{N}}/\eta_2 - \bar{\bar{\chi}}_{em} \cdot \bar{\bar{J}}/(2\eta_2) \\ \bar{\bar{\chi}}_{me}/2 + \bar{\bar{J}} \cdot \bar{\bar{N}} + \bar{\bar{\chi}}_{mm} \cdot \bar{\bar{J}}/(2\eta_1) & \bar{\bar{\chi}}_{me}/2 - \bar{\bar{J}} \cdot \bar{\bar{N}} - \bar{\bar{\chi}}_{mm} \cdot \bar{\bar{J}}/(2\eta_2) \end{pmatrix}. \tag{4.19}$$

Upon inspection of (4.16), an important result may be deduced, which are the general reflectionless conditions for a normally incident plane wave impinging on a uniform bianisotropic metasurface. In order to obtain such matching conditions, we simply set $\bar{\bar{S}}_{11} = 0$ and $\bar{\bar{S}}_{22} = 0$ in (4.16) and, assuming that $\eta_1 = \eta_2$, obtain the following relationships between the susceptibilities

$$\bar{\bar{\chi}}_{ee} = -\bar{\bar{J}} \cdot \bar{\bar{\chi}}_{mm} \cdot \bar{\bar{J}}, \tag{4.20a}$$

$$\bar{\bar{\chi}}_{em} = \bar{\bar{J}} \cdot \bar{\bar{\chi}}_{me} \cdot \bar{\bar{J}}. \tag{4.20b}$$

Note that for a reciprocal metasurface, for which conditions (2.52) are satisfied, relation (4.20b) reduces to

$$\bar{\bar{\chi}}_{em} = -\bar{\bar{\chi}}_{me}^T = \kappa \bar{\bar{I}}, \tag{4.21}$$

where κ is a coefficient related to chirality. This equality implies that a recipro-cal bianisotropic metasurface can be reflectionless only[2] when it corresponds to a chiral bi-isotropic structure [155].

For illustration, we now provide the expressions relating the susceptibili-ties to the scattering parameters (and vice versa) in the particular case of the homoanisotropic diagonal metasurface discussed in Section 4.1.2. We assume that the media on both sides of the metasurface are the same and correspond to vacuum, i.e. $\eta_1 = \eta_2 = \eta_0$. This metasurface is nongyrotropic and reciprocal. Therefore, we have $S_{ab}^{xy} = S_{ab}^{yx} = 0$ and $\overline{\overline{S}}_{21} = \overline{\overline{S}}_{12}^T$. Solving (4.11) along with (4.4) leads to the susceptibilities

$$\chi_{ee}^{xx} = \frac{2j\left(T_x + R_x - 1\right)}{k_0\left(T_x + R_x + 1\right)}, \tag{4.22a}$$

$$\chi_{ee}^{yy} = \frac{2j\left(T_y + R_y - 1\right)}{k_0\left(T_y + R_y + 1\right)}, \tag{4.22b}$$

$$\chi_{mm}^{xx} = \frac{2j\left(T_y - R_y - 1\right)}{k_0\left(T_y - R_y + 1\right)}, \tag{4.22c}$$

$$\chi_{mm}^{yy} = \frac{2j\left(T_x - R_x - 1\right)}{k_0\left(T_x - R_x + 1\right)}, \tag{4.22d}$$

where, for convenience, we have noted $T_x = S_{21}^{xx}$ and $R_x = S_{11}^{xx}$, and so on. Reversing these relations, so as to express the scattering parameters in terms of the suscepti-bilities, leads to

$$T_x = \frac{4 + \chi_{ee}^{xx}\chi_{mm}^{yy}k_0^2}{\left(2 + jk_0\chi_{ee}^{xx}\right)\left(2 + jk_0\chi_{mm}^{yy}\right)}, \tag{4.23a}$$

$$R_x = \frac{2jk_0\left(\chi_{mm}^{yy} - \chi_{ee}^{xx}\right)}{\left(2 + jk_0\chi_{ee}^{xx}\right)\left(2 + jk_0\chi_{mm}^{yy}\right)}, \tag{4.23b}$$

for x-polarized waves, and

$$T_y = \frac{4 + \chi_{ee}^{yy}\chi_{mm}^{xx}k_0^2}{\left(2 + jk_0\chi_{ee}^{yy}\right)\left(2 + jk_0\chi_{mm}^{xx}\right)}, \tag{4.24a}$$

$$R_y = \frac{2jk_0\left(\chi_{mm}^{xx} - \chi_{ee}^{yy}\right)}{\left(2 + jk_0\chi_{ee}^{yy}\right)\left(2 + jk_0\chi_{mm}^{xx}\right)}. \tag{4.24b}$$

---

2 These results are valid only for normally propagating waves interacting with a uniform metasurface surrounded by the same medium on both of its sides. For a nonuniform metasurface and/or a metasurface surrounded by different media, the reflectionless conditions are different.

for y-polarized waves. Note that to illustrate the relations between susceptibilities and scattering parameters, we will see several examples of their application in the forthcoming sections and chapters.

According to (4.23b) and (4.24b), the metasurface is reflectionless, or perfectly matched, if $\chi_{mm}^{yy} - \chi_{ee}^{xx}$ and $\chi_{mm}^{xx} = \chi_{ee}^{yy}$, which corresponds to a particular case of (4.20a). These equations essentially correspond to the Kerker conditions [78], according to which the forward or backward scattering may be canceled, providing proper equilibrium between electric and magnetic dipolar responses. For illustration, consider the superposition of an x-oriented electric dipole and a y-oriented magnetic dipole, shown in Figure 4.2, where the collocated dipoles mimic the dipolar properties of a metasurface scattering particle. The orientation of the electric and magnetic dipoles is such that their far-fields are opposite and in the same directions at the two sides of the dipole pair, and respectively sum up to zero and twice the field of each dipole if the amplitude and phase of the dipole sources are properly tuned. This results in the asymmetric radiation pattern, plotted on the right-hand side of Figure 4.2, which, in this case, clearly illustrates the cancellation of backward scattering.

At this point, one may wonder whether the scattering parameter expressions obtained above apply *only* to metasurfaces synthesized to transform incident, reflected, and transmitted waves that are normally propagating with respect to the metasurface, or whether they also apply for arbitrary field transformations, which generally involves nonuniform metasurfaces? And whether or not these scattering parameter expressions may generally be used to relate the shapes of the

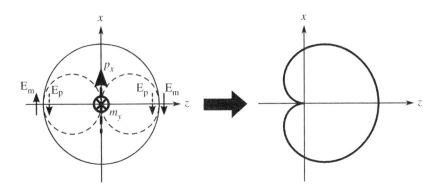

Figure 4.2 Illustration of the Kerker effect. An electric dipole, p_x, and a magnetic dipole, m_y, are collocated and their respective amplitude and phase are tuned so that the electric field of the electric dipole, $\mathbf{E}_p$ (dashed line), constructively or destructively interferes with the electric field of the magnetic dipole, $\mathbf{E}_m$ (solid line), leading to an asymmetric radiation pattern with canceled backward scattering.

scattering particles to the metasurface susceptibilities? To answer these questions, let us consider the following four cases: (i) the metasurface is synthesized only for normally propagating waves; (ii) the metasurface is synthesized for obliquely propagating waves but without changing the direction of wave propagation, i.e. the metasurface is uniform; (iii) the metasurface is synthesized to change the direction of wave propagation (e.g. refraction, collimation, etc.) but, at least, one of the specified wave propagates normally to the metasurface; and (iv) the metasurface is synthesized to change the direction of wave propagation but none of the specified waves propagates normally (e.g. negative refraction).

Case 1 The expressions derived above perfectly apply and the realized metasurface response will be in exact agreement with the specified one.

Case 2 For illustration, let us consider the synthesis of a reflectionless uniform metasurface that rotates the polarization of the incident wave impinging on the metasurface at an angle $\theta_i = \theta_t = 30°$ from broadside. If the expressions above are used and the scattering particle simulated with normal-wave propagation, then the response of the realized metasurface for the specified incidence angle of 30° will *not* corresponds to the expected result. Indeed, as mentioned above, simulating all the scattering parameters with normally propagating waves and solving (4.11) yields the exact tangential susceptibility components but does not provide any information about the normal susceptibility components. Therefore, an obliquely impinging wave may excite these normal polarizations, hence affecting the scattering response of the metasurface. The second reason is the non-negligible coupling between adjacent scattering particles, which depends on the angle of incidence of the excitation (spatial dispersion). This coupling is thus different for normal- and oblique-wave propagation. In order to correctly synthesize this metasurface, one should change relations (4.12) so as to include the angle of wave propagation. Accordingly, the scattering particles must be simulated with the same specified angle of wave propagation. Only then, would the metasurface yield the expected scattering response.

Case 3 For illustration, let us consider the synthesis of a reflectionless metasurface that refracts a normally incident plane wave at a given refraction angle θ_t. In this case, the metasurface is nonuniform and the susceptibilities are thus functions of x and y on the metasurface. Inserting these susceptibilities into the above relations yields scattering parameters that are themselves spatially varying. It is important to understand that these scattering parameters *do not* represent the overall scattering behavior of the metasurface but rather correspond to *local* scattering parameters. Due to the nonuniformity of the structure, several different scattering particles have to be realized, each one of them exhibiting the local scattering parameters corresponding to their position (x, y) on the metasurface. This is achieved by simulating each scattering particle *individually* with periodic boundary conditions (PBC). Consequently, once they are all implemented and combined together to form the final metasurface, the coupling between them differs from when they were simulated with PBC. This leads to a scattering response that is different from the expected one, irrespective of whether the relations above (assuming normal-wave propagation) are used or any other one (assuming oblique-wave propagation). In that case, the relations between the susceptibilities and the scattering parameters only serve as an initial guess to realize the scattering particles, which then usually require to be optimized so that the final metasurface performs the expected response. In addition to this, the realized scattering particles may also exhibit nonzero normal susceptibilities, which further degrades the scattering behavior of the metasurface. Since, in the example that is considered here, the incident wave is normally impinging on the metasurface, the relations derived above provide a relatively good initial guess to realize the metasurface scattering particles because the excitation of the normal susceptibility components is limited.

Case 4 This case is a combination of the two previous ones since all the waves propagate obliquely (and at different angles) with respect to the metasurface that is thus nonuniform. If the specified incident, reflected, and transmitted fields are all plane waves, then it is adequate to adjust the direction of wave propagation to that of the specified incident wave in (4.12) as well as in the full-wave simulations. In contrast, if the specified incident, reflected, and transmitted fields are not plane waves, but arbitrary fields, then the relations derived above are sufficient to obtain an initial approximation of the specified scattering behavior that may require further optimization. This is because metasurfaces are usually synthesized by assuming that the normal susceptibility components are zero. Therefore, simulating the scattering particles with normally propagating waves ensures that their resulting tangential susceptibilities correspond to the desired ones obtained from the synthesis. Then, the discrepancies in the final metasurface response may only come from the presence of nonzero normal polarizations and coupling with adjacent unit cells. However, at least their tangential polarizations are the expected ones, which minimizes the errors in the scattering behavior of the metasurface.

In general, the susceptibilities obtained from the synthesis are complex quantities, which may be related to loss and gain. Specifically, the susceptibilities in (4.7) correspond to dissipation if their imaginary parts are negative, and gain if their imaginary parts are positive. This is relevant to the implementation of metasurfaces because the realization of the scattering particles is generally simplified for structures made of purely lossless dielectric material and metallic inclusions made of perfect electric conductor (PEC). In this case, the susceptibilities retrieved with (4.22) are necessarily purely real. Consequently, it would be impossible to realize the exact complex susceptibilities obtained from the synthesis with such kind of "ideal" scattering particles.

There are, at least, three strategies to overcome this issue. The first consists in using lossy materials to approach, as closely as possible, the required susceptibility values. This is generally difficult for it would require a precise control of the dissipation as function of the position on the metasurface. The second strategy consists in simply setting to zero the imaginary parts of the synthesized susceptibilities and implementing only their remaining real parts. In this case, the metasurface response will differ from the expected one as undesired scattering will appear, which, depending on the application, may or may not be an issue. The third strategy is similar to the second one, in the sense that the imaginary parts of the synthesized susceptibilities are set to zero, but the remaining real parts are optimized so as to minimize the discrepancies between the expected response of the metasurface and the approximated one. Typically, the optimization procedure consists in minimizing a cost function of the following form [125]:

$$F = \left| T_{\text{spec}} - T_{\text{approx}} \right|^2 + \left| R_{\text{spec}} - R_{\text{approx}} \right|^2, \qquad (4.25)$$

where the scattering parameters T_{spec} and R_{spec} are obtained from (4.23) or (4.24) using the synthesized susceptibilities. While the parameters T_{approx} and R_{approx} are obtained from the same equations but with the purely real susceptibilities that have to be optimized. The cost function (4.25) is naturally extendable to the more general case of full scattering parameter matrices.

4.1.5 Surface-Wave Eigenvalue Problem

4.1.5.1 Formulation of the Problem

In the previous section, we have seen how the GSTCs may be used to model and synthesize metasurfaces performing essentially far-field transformations, and we have not yet discussed the case of surface-wave propagation. From a theoretical perspective, the GSTCs are perfectly capable to handle surface-wave and near-field transformations. For the specific case of surface-wave propagation, it is possible to transform the GSTC relations (4.2) into an eigenvalue problem, which may be used to either obtain the dispersion diagram of the metasurface, assuming known susceptibilities, or to synthesize the metasurface susceptibilities so as to achieve a desired surface-wave propagation with specific propagation characteristics [3].

We shall now start by showing how to transform the GSTC relations (4.2) into an eigenvalue problem relating the metasurface susceptibilities to the wavenumbers of surface waves. For this purpose, consider the geometry of Figure 4.3, where a *uniform* metasurface lies in the xy-plane between two different media.

For simplicity but without loss of generality, we next restrict our attention to the case of surface-wave propagation in the x-direction. We also consider that the metasurface may support both TE and TM surface-wave modes. In order to relate the metasurface susceptibilities to the propagation characteristics of surface waves, we express the electromagnetic fields at the top ($z = 0^+$) and bottom ($z = 0^-$) sides of the metasurface as the superposition of TE and TM surface

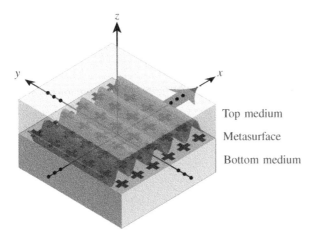

Figure 4.3 Surface wave propagating on a metasurface sandwiched between two different media [3].

waves. Since we are interested in formulating an eigenvalue problem, the incident field is set to zero. Accordingly, the fields at the bottom side may be written

$$\mathbf{E}^{0^-} = \left(-\hat{\mathbf{x}} \frac{k_{1z}}{k_1} \eta_1 A_{\mathrm{TM}}^{0^-} + \hat{\mathbf{y}} A_{\mathrm{TE}}^{0^-} \right) \mathrm{e}^{-jk_x x}, \tag{4.26a}$$

$$\mathbf{H}^{0^-} = \left(\hat{\mathbf{x}} \frac{k_{1z}}{k_1 \eta_1} A_{\mathrm{TE}}^{0^-} + \hat{\mathbf{y}} A_{\mathrm{TM}}^{0^-} \right) \mathrm{e}^{-jk_x x}, \tag{4.26b}$$

while those at the top side may be written

$$\mathbf{E}^{0^+} = \left(\hat{\mathbf{x}} \frac{k_{2z}}{k_2} \eta_2 A_{\mathrm{TM}}^{0^+} + \hat{\mathbf{y}} A_{\mathrm{TE}}^{0^+} \right) \mathrm{e}^{-jk_x x}, \tag{4.27a}$$

$$\mathbf{H}^{0^+} = \left(-\hat{\mathbf{x}} \frac{k_{2z}}{k_2 \eta_2} A_{\mathrm{TE}}^{0^+} + \hat{\mathbf{y}} A_{\mathrm{TM}}^{0^+} \right) \mathrm{e}^{-jk_x x}, \tag{4.27b}$$

where the parameters A are the complex amplitudes of the TE and TM surface waves at the top and bottom sides and $k_a^2 = k_{az}^2 + k_x^2$ with $a = 1$ or $a = 2$ correspond to the bottom and top sides' propagation constants, respectively. Note that, by phase matching, the top and bottom surface waves share the same tangential wavenumber, k_x, since the metasurface is uniform, and hence cannot induce any discontinuity in the tangential components of the fields.

Now, we substitute (4.26) and (4.27) into the GSTC relations (4.2) using the definition of the average and difference of the fields. After simplifications, we arrive at the following system of equations:

$$\overline{\overline{\chi}} \cdot \mathbf{x} = 0, \tag{4.28}$$

where **x** is a vector containing the amplitude of the surface-wave modes and reads

$$\mathbf{x}^{\mathrm{T}} = \left(A_{\mathrm{TE}}^{0-}, A_{\mathrm{TM}}^{0-}, A_{\mathrm{TE}}^{0+}, A_{\mathrm{TM}}^{0+} \right), \tag{4.29}$$

and $\overline{\overline{\chi}}$ is a matrix containing the metasurface susceptibilities and the surface waves propagation constants, which is given by

$$\overline{\overline{\chi}} = \begin{pmatrix} \overline{\overline{\chi}}_{11} & \overline{\overline{\chi}}_{12} \\ \overline{\overline{\chi}}_{21} & \overline{\overline{\chi}}_{22} \end{pmatrix}, \tag{4.30}$$

with the sub-matrices being

$$\overline{\overline{\chi}}_{11} = \begin{pmatrix} j\omega\epsilon\,\chi_{\mathrm{ee}}^{xy} + \dfrac{jk_{1z}k\,\chi_{\mathrm{em}}^{xx}}{k_1\eta_1} & -2 + jk\chi_{\mathrm{em}}^{xy} - \dfrac{jk_{1z}\eta_1\omega\epsilon\,\chi_{\mathrm{ee}}^{xx}}{k_1} \\ j\omega\epsilon\,\chi_{\mathrm{ee}}^{yy} + \dfrac{k_{1z}\left(2 + jk\chi_{\mathrm{em}}^{yx}\right)}{k_1\eta_1} & jk\chi_{\mathrm{em}}^{yy} - \dfrac{jk_{1z}\eta_1^2\omega\epsilon\,\chi_{\mathrm{ee}}^{yx}}{k_1\eta_1} \end{pmatrix}, \tag{4.31a}$$

$$\overline{\overline{\chi}}_{12} = \begin{pmatrix} j\omega\epsilon\,\chi_{\mathrm{ee}}^{xy} - \dfrac{jk_{2z}k\,\chi_{\mathrm{em}}^{xx}}{k_2\eta_2} & 2 + jk\chi_{\mathrm{em}}^{xy} + \dfrac{jk_{2z}\eta_2\omega\epsilon\,\chi_{\mathrm{ee}}^{xx}}{k_2} \\ j\omega\epsilon\,\chi_{\mathrm{ee}}^{yy} + \dfrac{k_{2z}\left(2 - jk\chi_{\mathrm{em}}^{yx}\right)}{k_2\eta_2} & jk\chi_{\mathrm{em}}^{yy} + \dfrac{jk_{2z}\eta_2^2\omega\epsilon\,\chi_{\mathrm{ee}}^{yx}}{k_2\eta_2} \end{pmatrix}, \tag{4.31b}$$

$$\overline{\overline{\chi}}_{21} = \begin{pmatrix} -2 - jk\chi_{\mathrm{me}}^{xy} - \dfrac{jk_{1z}\omega\mu\,\chi_{\mathrm{mm}}^{xx}}{k_1\eta_1} & -j\omega\mu\chi_{\mathrm{mm}}^{xy} + \dfrac{jk_{1z}\eta_1 k\,\chi_{\mathrm{me}}^{xx}}{k_1} \\ -jk\chi_{\mathrm{me}}^{yy} - \dfrac{jk_{1z}\omega\mu\,\chi_{\mathrm{mm}}^{yx}}{k_1\eta_1} & -j\omega\mu\chi_{\mathrm{mm}}^{yy} - \dfrac{k_{1z}\eta_1\left(2 - jk\chi_{\mathrm{me}}^{yx}\right)}{k_1} \end{pmatrix}, \tag{4.31c}$$

$$\overline{\overline{\chi}}_{22} = \begin{pmatrix} 2 - jk\chi_{\mathrm{me}}^{xy} + \dfrac{jk_{2z}\omega\mu\,\chi_{\mathrm{mm}}^{xx}}{k_2\eta_2} & -j\omega\mu\chi_{\mathrm{mm}}^{xy} - \dfrac{jk_{2z}\eta_2 k\,\chi_{\mathrm{me}}^{xx}}{k_2} \\ -jk\chi_{\mathrm{me}}^{yy} + \dfrac{jk_{2z}\omega\mu\,\chi_{\mathrm{mm}}^{yx}}{k_2\eta_2} & -j\omega\mu\chi_{\mathrm{mm}}^{yy} - \dfrac{k_{2z}\eta_2\left(2 + jk\chi_{\mathrm{me}}^{yx}\right)}{k_2} \end{pmatrix}. \tag{4.31d}$$

The system (4.28) provides the sought-after relationships between susceptibilities and the surface-wave characteristics. It may hence be solved to express either one in terms of the other.

The propagation constant of a surface wave propagating on a metasurface with known susceptibilities and embedded in an asymmetric environment, i.e. when the media on both sides of the metasurface are different, may be computed by replacing k_{az} in (4.28) by $\pm\sqrt{k_a^2 - k_x^2}$, where $a = \{1, 2\}$. This leads to a nonlinear system of equations, which may be solved (numerically) for the propagation constant k_x. We let the reader discover more about the propagation of surface waves in asymmetric environment in [3] and shall now rather discuss the case of dispersion in symmetric environments.

4.1.5.2 Dispersion in a Symmetric Environment

In a symmetric environment, the media at both sides of the metasurface are the same, i.e. $\eta_1 = \eta_2 = \eta$, $k_1 = k_2 = k$ and, by consequence of phase matching, $k_{1z} = k_{2z} = k_z$ since $k_{1x} = k_{2x} = k_x$. By extracting the longitudinal wavenumber k_z from (4.30), we can transform (4.28) into

$$\overline{\overline{A}} \cdot \mathbf{x} = k_z \overline{\overline{B}} \cdot \mathbf{x}, \tag{4.32}$$

where the matrices $\overline{\overline{A}}$ and $\overline{\overline{B}}$ are given by

$$\overline{\overline{A}} = \begin{pmatrix} -k\chi_{ee}^{xy} & -2j\eta - k\eta\chi_{em}^{xy} & -k\chi_{ee}^{xy} & 2j\eta - k\eta\chi_{em}^{xy} \\ k^2\chi_{ee}^{yy} & k^2\eta\chi_{em}^{yy} & k^2\chi_{ee}^{yy} & k^2\eta\chi_{em}^{yy} \\ -2j + k\chi_{me}^{xy} & k\eta\chi_{mm}^{xy} & 2j + k\chi_{me}^{xy} & k\eta\chi_{mm}^{xy} \\ k^2\chi_{me}^{yy} & k^2\eta\chi_{mm}^{yy} & k^2\chi_{me}^{yy} & k^2\eta\chi_{mm}^{yy} \end{pmatrix}, \tag{4.33a}$$

$$\overline{\overline{B}} = \begin{pmatrix} \chi_{em}^{xx} & -\eta\chi_{ee}^{xx} & -\chi_{em}^{xx} & \eta\chi_{ee}^{xx} \\ 2j - k\chi_{em}^{yx} & k\eta\chi_{ee}^{yx} & 2j + k\chi_{em}^{yx} & -k\eta\chi_{ee}^{yx} \\ -\chi_{mm}^{xx} & \eta\chi_{me}^{xx} & \chi_{mm}^{xx} & -\eta\chi_{me}^{xx} \\ -k\chi_{mm}^{yx} & 2j\eta + k\eta\chi_{me}^{yx} & k\chi_{mm}^{yx} & 2j\eta - k\eta\chi_{me}^{yx} \end{pmatrix}. \tag{4.33b}$$

Pre-multiplying (4.32) by the inverse of $\overline{\overline{B}}$ yields the eigenvalue equation

$$\overline{\overline{M}} \cdot \mathbf{x} = k_z \mathbf{x}, \tag{4.34}$$

where $\overline{\overline{M}} = \overline{\overline{B}}^{-1} \cdot \overline{\overline{A}}$, and where k_z and $\mathbf{x}$ are the eigenvalue and eigenvector of the problem, respectively.

Next, we solve (4.34) to provide the relationships between susceptibilities and surface-wave characteristics in the case of a birefringent metasurface, for which only the susceptibility components χ_{ee}^{xx}, χ_{ee}^{yy}, χ_{mm}^{xx} and χ_{mm}^{yy} are nonzero. The resulting relationships are provided in Table 4.1. Note that the absolute value of the amplitude of the surface waves cannot be found directly from (4.34). Nevertheless, we may still obtain the relative amplitude between the top and bottom waves.

From Table 4.1, we see that the TE and TM modes each split into symmetric and asymmetric field configurations, where the symmetric ones are associated with a lower frequency (ω^-) than the asymmetric ones (ω^+). Here the notion of symmetric and asymmetric fields refers to the tangential component of the electric field, E_x. Such field distributions are illustrated in Figure 4.4 for surface waves propagating on a 60 nm thick silver slab at $\lambda = 400$ nm and surrounded by vacuum. We consider here the case of a slab instead of a metasurface in order to validate our model with an actual structure.

Table 4.1 may be used either to obtain the dispersion diagram of a metasurface with known susceptibilities or to synthesize the metasurface for the propagation

Table 4.1 Dispersion relations of birefringent metasurfaces.

Eigenvalues, k_z	Eigenvectors, $\mathbf{x}^\mathsf{T}$	Wavenumbers, k_x
$\dfrac{2j}{\chi_{\mathrm{mm}}^{xx}}$	$(-1,0,1,0)$ TE, symmetric, ω^-	$\pm\dfrac{\sqrt{4+(k\chi_{\mathrm{mm}}^{xx})^2}}{\chi_{\mathrm{mm}}^{xx}}$
$\dfrac{2j}{\chi_{\mathrm{ee}}^{xx}}$	$(0,-1,0,1)$ TM, symmetric, ω^-	$\pm\dfrac{\sqrt{4+\left(k\chi_{\mathrm{ee}}^{xx}\right)^2}}{\chi_{\mathrm{ee}}^{xx}}$
$-\dfrac{1}{2}jk^2\chi_{\mathrm{ee}}^{yy}$	$(1,0,1,0)$ TE, asymmetric, ω^+	$\pm\dfrac{1}{2}k\sqrt{4+\left(k\chi_{\mathrm{ee}}^{yy}\right)^2}$
$-\dfrac{1}{2}jk^2\chi_{\mathrm{mm}}^{yy}$	$(0,1,0,1)$ TM, asymmetric, ω^+	$\pm\dfrac{1}{2}k\sqrt{4+\left(k\chi_{\mathrm{mm}}^{yy}\right)^2}$

of a surface wave with specified propagation constant. The latter operation is achieved by specifying k_x, in the third column of Table 4.1, and then solving the corresponding expression for the susceptibilities. The former operation requires an a priori knowledge of the metasurface susceptibilities in order to compute the corresponding propagation constant. If the susceptibilities cannot be found analytically, they can still be obtained numerically by computing the metasurface scattering parameters using full-wave simulations, and then relating them to the effective susceptibilities, as explained in Section 4.1.4. This is illustrated by computing the dispersion of the silver slab shown in Figure 4.4. The dispersion curves of the symmetric and asymmetric TM surface wave modes of this slab are respectively plotted in Figure 4.5, where the dashed curves correspond to the

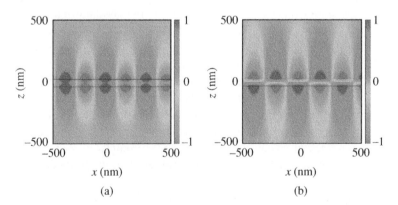

Figure 4.4 Simulated real part of E_x for surface waves, with (a) symmetric and (b) asymmetric field distributions [3].

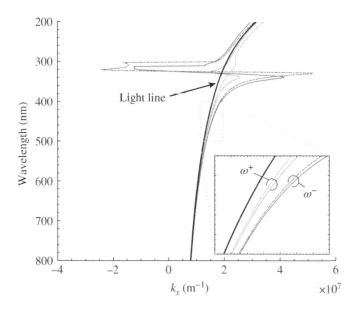

Figure 4.5 Dispersion of a 60 nm thick silver slab [3].

GSTC method,[3] while the solid curves correspond to the exact solutions from Maxwell equations [138].

To further illustrate the application of (4.34), we now consider the case of a reciprocal Ω-type bianisotropic metasurface, for which the nonzero susceptibility components are χ_{ee}^{xx}, χ_{ee}^{yy}, χ_{mm}^{xx}, χ_{mm}^{yy}, χ_{em}^{xy}, χ_{em}^{yx}, and χ_{me}^{xy}, χ_{me}^{yx}. The relationships between susceptibilities and surface-wave characteristics are provided in Table 4.2, where the wavenumbers k_x are easily calculated by solving $k^2 = k_z^2 + k_x^2$. In addition to the extra complexity, the main difference between a birefringent and an Ω-type bianisotropic metasurface resides in the ability of the latter to control the ratio between the top and bottom surface waves. Indeed, as evidenced by the presence of the susceptibilities in the eigenvectors **x**, changing the values of these susceptibilities allows one to control the amplitude of the surface waves. This means that such a metasurface has the inherent capability to create strongly asymmetric surface waves and even completely prevent surface-wave propagation at one side of it [3].

4.1.6 Metasurfaces with Normal Polarizations

The metasurface synthesis relations derived in Section 4.1 have been obtained under the assumption that the metasurface does not possess normal susceptibility

---

3 Here, the susceptibilities of the slab are obtained by converting the scattering parameters of the slab for a normally incident plane wave into effective susceptibilities following the explanation in Section 4.1.4.

Table 4.2 Dispersion relations for omega-type metasurfaces.

Eigenvalues, k_z	Eigenvectors, $\mathbf{x}^{\mathsf{T}}$
$\dfrac{k^2\left(\chi_{\text{em}}^{2yx} + \chi_{\text{ee}}^{yy}\chi_{\text{mm}}^{xx}\right) - 4 + C_{\text{TE}}}{4j\chi_{\text{mm}}^{xx}}$	$\left(\dfrac{4jk\chi_{\text{em}}^{yx} + C_{\text{TE}}}{k^2\left(\chi_{\text{em}}^{2yx} + \chi_{\text{ee}}^{yy}\chi_{\text{mm}}^{xx}\right) + 4}, 0, 1, 0 \right)$
$\dfrac{k^2\left(\chi_{\text{em}}^{2yx} + \chi_{\text{ee}}^{yy}\chi_{\text{mm}}^{xx}\right) - 4 - C_{\text{TE}}}{4j\chi_{\text{mm}}^{xx}}$	$\left(\dfrac{4jk\chi_{\text{em}}^{yx} - C_{\text{TE}}}{k^2\left(\chi_{\text{em}}^{2yx} + \chi_{\text{ee}}^{yy}\chi_{\text{mm}}^{xx}\right) + 4}, 0, 1, 0 \right)$
$\dfrac{k^2\left(\chi_{\text{em}}^{2xy} + \chi_{\text{ee}}^{xx}\chi_{\text{mm}}^{yy}\right) - 4 + C_{\text{TM}}}{4j\chi_{\text{ee}}^{xx}}$	$\left(0, \dfrac{4jk\chi_{\text{em}}^{xy} + C_{\text{TM}}}{k^2\left(\chi_{\text{em}}^{2xy} + \chi_{\text{ee}}^{xx}\chi_{\text{mm}}^{yy}\right) + 4}, 0, 1 \right)$
$\dfrac{k^2\left(\chi_{\text{em}}^{2xy} + \chi_{\text{ee}}^{xx}\chi_{\text{mm}}^{yy}\right) - 4 - C_{\text{TM}}}{4j\chi_{\text{ee}}^{xx}}$	$\left(0, \dfrac{4jk\chi_{\text{em}}^{xy} - C_{\text{TM}}}{k^2\left(\chi_{\text{em}}^{2xy} + \chi_{\text{ee}}^{xx}\chi_{\text{mm}}^{yy}\right) + 4}, 0, 1 \right)$

$$C_{\text{TE}} = \sqrt{k^4\chi_{\text{em}}^{4yx} + 2k^2\chi_{\text{em}}^{2yx}(k^2\chi_{\text{ee}}^{yy}\chi_{\text{mm}}^{xx} - 4) + (k^2\chi_{\text{ee}}^{yy}\chi_{\text{mm}}^{xx} + 4)^2}$$

$$C_{\text{TM}} = \sqrt{k^4\chi_{\text{em}}^{4xy} + 2k^2\chi_{\text{em}}^{2xy}(k^2\chi_{\text{ee}}^{xx}\chi_{\text{mm}}^{yy} - 4) + (k^2\chi_{\text{ee}}^{xx}\chi_{\text{mm}}^{yy} + 4)^2}$$

components. This restriction was used to simplify the GSTCs since the spatial derivatives in (4.1) vanish in the absence of normal polarization densities. For completeness, we shall now discuss metasurfaces with nonzero normal polarization densities.

The general bianisotropic system of equations (4.1) involves a total number of 36 unknown susceptibilities for only 4 equations when full susceptibility tensors are considered, instead of 16 unknown as, for instance, considered in Section 4.1.4. This corresponds to a heavily underdetermined system, where these additional susceptibility components largely increase the number of degrees of freedom available to control electromagnetic waves.

The synthesis of a metasurface with nonzero normal polarization densities may be performed following procedures similar to those discussed in Section 4.1. One needs to equate the number of unknown susceptibilities to the number of equations provided by the GSTCs. However, this may become more challenging when considering the presence of normal susceptibility components since more transformations may be required to obtain a full-rank system. Moreover, if the specified transformations involve changing the direction of wave propagation, then the system (4.1) becomes a coupled system of partial differential equations due to the nonvanishing presence of the spatial derivatives. In general, this prevents the derivation of closed-form solutions for the susceptibilities, which may therefore have to be obtained numerically.

The relations between the susceptibilities and scattering parameters can also be obtained by the procedure presented in Section 4.1.4. If the susceptibilities spatially vary in the plane of the metasurface, the scattering parameters are obtained by solving relations (4.1), which, as mentioned above, form a set of coupled differential equations that must be solved numerically. However, if the metasurface susceptibility functions do not vary spatially, Snell's law applies and the scattering parameters can thus be obtained in closed form. In that case, it is possible to derive relations similar to (4.13) and (4.14), and (4.18) and (4.19), that apply in the most general case where the 36 susceptibilities are considered. In Section 4.1.4, we expressed the 16 susceptibilities in terms of scattering parameters by considering the 4 GSTC equations along with 4 wave transformations: x- and y-polarized excitations from port 1 and x- and y-polarized excitations from port 2. Now, if the 36 susceptibilities are considered, we would need to specify nine independent wave transformations so as to form a full-rank system with the same four GSTC equations. Although feasible, it is particularly tedious and thus avoided here. Alternatively, we can express the scattering parameters as functions of the susceptibilities. This is easier because the system of equations is always a full-rank system made of 16 unknown scattering parameters distributed among 4 wave transformations, as was the case for the derivation of (4.18) and (4.19). Unfortunately, although expressing the scattering parameters in terms of the susceptibilities is easier, the resulting relations are very lengthy and are therefore not given here.

Instead, we illustrate this procedure by providing the expressions of the reflection and transmission coefficients for the particular case of a homoanisotropic uniform metasurface with diagonal susceptibility tensors [56]. Its susceptibility tensors are

$$\overline{\overline{\chi}}_{ee} = \begin{pmatrix} \chi_{ee}^{xx} & 0 & 0 \\ 0 & \chi_{ee}^{yy} & 0 \\ 0 & 0 & \chi_{ee}^{zz} \end{pmatrix}, \quad \overline{\overline{\chi}}_{mm} = \begin{pmatrix} \chi_{mm}^{xx} & 0 & 0 \\ 0 & \chi_{mm}^{yy} & 0 \\ 0 & 0 & \chi_{mm}^{zz} \end{pmatrix}. \quad (4.35)$$

We illuminate the metasurface so that scattering occurs only in the xz-plane. Inserting (4.35) into (4.1) leads to the reduced GSTC relations

$$-\Delta H_y = j\omega\epsilon_0 \chi_{ee}^{xx} E_x, \quad (4.36a)$$

$$\Delta H_x = j\omega\epsilon_0 \chi_{ee}^{yy} E_y - \chi_{mm}^{zz} \partial_x H_z, \quad (4.36b)$$

$$\Delta E_y = j\omega\mu_0 \chi_{mm}^{xx} H_x, \quad (4.36c)$$

$$-\Delta E_x = j\omega\mu_0 \chi_{mm}^{yy} H_y + \chi_{ee}^{zz} \partial_x E_z. \quad (4.36d)$$

Note that the partial derivatives in (4.36b) and (4.36d) are only in the x-direction since the scattering from this metasurface is restricted to the xz-plane, and that they only apply to the fields and not to the susceptibilities since the latter are spatially uniform.

If such a metasurface were illuminated by a normally incident plane wave, then the susceptibilities χ_{ee}^{zz} and χ_{mm}^{zz} would not be excited, since $E_z = H_z = 0$ for normal propagation. In that case, the scattering parameters are exactly identical to those in (4.23) and (4.24). The situation is more interesting in the case of oblique incidence. Given that $\partial_x \to -jk_x$ and using relations similar to (4.12) for the specification of the fields, the scattering parameters for p-polarized waves are found as

$$T_p = \frac{k_z(4 + \chi_{ee}^{xx}\chi_{mm}^{yy}k_0^2 + \chi_{ee}^{xx}\chi_{ee}^{zz}k_x^2)}{(2j - \chi_{ee}^{xx}k_z)(\chi_{mm}^{yy}k_0^2 + \chi_{ee}^{zz}k_x^2 - 2jk_z)}, \tag{4.37a}$$

$$R_p = \frac{2j(\chi_{mm}^{yy}k_0^2 + \chi_{ee}^{zz}k_x^2 - \chi_{ee}^{xx}k_z^2)}{(2j - \chi_{ee}^{xx}k_z)(\chi_{mm}^{yy}k_0^2 + \chi_{ee}^{zz}k_x^2 - 2jk_z)}, \tag{4.37b}$$

and their counterparts for s-polarized waves are

$$T_s = \frac{k_z(4 + \chi_{mm}^{xx}\chi_{ee}^{yy}k_0^2 + \chi_{mm}^{xx}\chi_{mm}^{zz}k_x^2)}{(2j - \chi_{mm}^{xx}k_z)(\chi_{ee}^{yy}k_0^2 + \chi_{mm}^{zz}k_x^2 - 2jk_z)}, \tag{4.38a}$$

$$R_s = -\frac{2j(\chi_{ee}^{yy}k_0^2 + \chi_{mm}^{zz}k_x^2 - \chi_{mm}^{xx}k_z^2)}{(2j - \chi_{mm}^{xx}k_z)(\chi_{ee}^{yy}k_0^2 + \chi_{mm}^{zz}k_x^2 - 2jk_z)}. \tag{4.38b}$$

It can be easily verified that, for normal incidence where $k_z = k_0$ and $k_x = 0$, the relations (4.37) and (4.38) respectively reduce to (4.23) and (4.24). Alternatively, by setting $\chi_{ee}^{zz} = \chi_{mm}^{zz} = 0$, the relations (4.37) and (4.38) may be used to generalize the relations (4.23) and (4.24) with the reflection and transmission coefficients of a uniform metasurface expressed as functions of the incidence angle.

4.1.7 Illustrative Examples

We will now illustrate the metasurface synthesis procedure with three different examples. The first one presents different combinations of susceptibilities performing the same operation of polarization rotation, as an illustration of the single transformation technique described in Section 4.1.2 and the scattering parameters technique presented in Section 4.1.4. The second example illustrates the application of the multiple transformation technique discussed in Section 4.1.3 for the synthesis of a nonreciprocal metasurface. Finally, the third example combines the concept of multiple transformations with nonzero normal susceptibility components to realize a reciprocal metasurface with controllable angular scattering.

4.1.7.1 Polarization Rotation

We shall now consider a metasurface synthesis example that illustrates the application of the techniques discussed in Sections 4.1.2 and 4.1.4. Let us consider the following synthesis problem: find the susceptibilities of a reflectionless metasurface that rotates the polarization of a normally incident plane wave by an angle of $\pi/3$. We specify that the incident wave is linearly polarized with an electric field making a $\pi/8$ angle with respect to the x-axis, as shown in Figure 4.6. The resulting transmitted wave is thus polarized with an electric field making an angle of $11\pi/24$. We assume that the metasurface is surrounded on both sides by vacuum, i.e. $\eta_1 = \eta_2 = \eta_0$.

The incident and transmitted electromagnetic fields corresponding to these specifications, respectively, read at $z = 0$

$$\mathbf{E}_i(x,y) = \hat{\mathbf{x}}\cos(\pi/8) + \hat{\mathbf{y}}\sin(\pi/8), \tag{4.39a}$$

$$\mathbf{H}_i(x,y) = \frac{1}{\eta_0}\left[-\hat{\mathbf{x}}\sin(\pi/8) + \hat{\mathbf{y}}\cos(\pi/8)\right], \tag{4.39b}$$

and

$$\mathbf{E}_t(x,y) = \hat{\mathbf{x}}\cos(11\pi/24) + \hat{\mathbf{y}}\sin(11\pi/24), \tag{4.40a}$$

$$\mathbf{H}_t(x,y) = \frac{1}{\eta_0}\left[-\hat{\mathbf{x}}\sin(11\pi/24) + \hat{\mathbf{y}}\cos(11\pi/24)\right]. \tag{4.40b}$$

We shall now consider four different sets of susceptibilities to realize these specifications to illustrate the diversity of possible designs for a given transformation.

The first synthesis assumes a birefringent metasurface, with susceptibilities given in (4.7). From the definition of the difference and average fields, the

Figure 4.6 The polarization of a normally incident plane wave, linearly polarized at a $\pi/8$ angle with respect to the x-axis, is rotated by $\pi/3$. In the reciprocal case, the fields retrieve their initial polarization upon propagation along the negative z-direction, while in the nonreciprocal case the fields rotate by a total of $2\pi/3$.

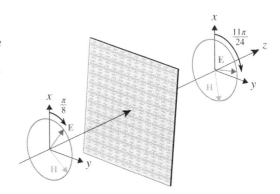

susceptibilities are readily obtained as

$$\chi_{ee}^{xx} = \chi_{mm}^{yy} = -\frac{1.5048}{k_0}j, \qquad (4.41a)$$

$$\chi_{ee}^{yy} = \chi_{mm}^{xx} = \frac{0.88603}{k_0}j. \qquad (4.41b)$$

The susceptibilities are spatially uniform, since the specified transformation does not alter the direction of wave propagation. Moreover, we know from the discussion in Section 4.1.2 and from relations (2.52) that the susceptibilities in (4.41) correspond to a reciprocal metasurface. However, we also know from (2.5) that, for diagonal susceptibilities such as (4.41), the negative imaginary parts correspond to absorption while positive imaginary parts correspond to gain. As a result, although the synthesized metasurface is *not* gyrotropic, since only the diagonal components of the monoanisotropic susceptibility tensors are nonzero, it is still able to perform the specified rotation of polarization by reducing the amplitudes of $E_{x,i}$ and $H_{y,i}$ while amplifying the amplitudes of $E_{y,i}$ and $H_{x,i}$. Moreover, this metasurface is *only* able to rotate the polarization angle by $\pi/3$ when the incident wave is polarized at a $\pi/8$ angle. If, for instance, the incident wave was only polarized along x, then only the susceptibilities in (4.41a) would be excited and the resulting transmitted field would still be polarized along x but with a reduced amplitude with respect to that of the incident wave due to the loss induced by these susceptibilities. Such a system is called a wave plate [143] and, since it does not rotate the polarization for arbitrary incidence, it has limited applications.

We shall now synthesize the same rotation of polarization transformation with an actual gyrotropic metasurface. For this purpose, we leverage the off-diagonal elements of a monoanisotropic metasurface. Solving (4.3) by setting all the susceptibilities to zero except χ_{ee}^{xy}, χ_{ee}^{yx}, χ_{mm}^{xy}, and χ_{mm}^{yx} yields the closed-form relations

$$\chi_{ee}^{xy} = \frac{-\Delta H_y}{j\omega\epsilon_0 E_{y,av}}, \qquad (4.42a)$$

$$\chi_{ee}^{yx} = \frac{\Delta H_x}{j\omega\epsilon_0 E_{x,av}}, \qquad (4.42b)$$

$$\chi_{mm}^{xy} = \frac{\Delta E_y}{j\omega\mu_0 H_{y,av}}, \qquad (4.42c)$$

$$\chi_{mm}^{yx} = \frac{-\Delta E_x}{j\omega\mu_0 H_{x,av}}, \qquad (4.42d)$$

which, upon substitution of the fields in (4.39) and (4.40), reduce to

$$\chi_{ee}^{xy} = \chi_{mm}^{xy} = -\frac{1.1547}{k_0}j, \tag{4.43a}$$

$$\chi_{ee}^{yx} = \chi_{mm}^{yx} = \frac{1.1547}{k_0}j. \tag{4.43b}$$

Contrary to the susceptibilities in (4.41), those in (4.43) perform a $\pi/3$-rotation of polarization *irrespectively* of the initial polarization of the incident wave, due to the truly gyrotropic nature of the metasurface. Additionally, the susceptibilities in (4.43) correspond to a nonreciprocal metasurface, given the relation of the reciprocity conditions in (2.52). Therefore, the metasurface is a Faraday rotating surface [80, 158], which exhibits the same polarization rotation direction irrespective of the direction of wave propagation. However, contrary to conventional Faraday rotators [81], this metasurface is also reflectionless due to the presence of both electric and magnetic gyrotropic susceptibility components, corresponding to rotating Huygens sources [113]. Interestingly, we can see that this metasurface is also gainless and lossless by applying relations (2.77). This metasurface is more general than the previously synthesized one, but it requires the implementation of nonreciprocal electric and magnetic materials, which may not be trivial to realize.

Still, following the same procedure, we now synthesize the polarization rotating metasurface using the off-diagonal hetero susceptibilities in (4.3). This yields the nonzero susceptibilities

$$\chi_{em}^{xy} = \chi_{me}^{yx} = -\frac{1.15048}{k_0}j, \tag{4.44a}$$

$$\chi_{em}^{yx} = \chi_{me}^{xy} = -\frac{0.88603}{k_0}j. \tag{4.44b}$$

This solution, despite being electromagnetically valid, turns out to be practically the most problematic so far. Indeed, such susceptibilities correspond to a nonreciprocal, active, and lossy metasurface, which performs the desired polarization transformation only for the specified incident wave. Such a metasurface may be shown to be equivalent to a "moving" medium [148].

Finally, we now consider the last four set of susceptibilities in (4.3). Instead of following the synthesis procedure used in the previous three cases, we shall apply the scattering-parameter synthesis method discussed in Section 4.1.4. In this case, we do not specify the transformation in terms of fields but rather using scattering parameters. The utilization of the scattering parameters is here justified by the fact that all the waves are propagating normally to the metasurface, which exactly corresponds to the prescription of the scattering parameters in (4.12). We assign ports 1 and 2 to the left- and right-hand sides of the metasurface, respectively. Since the metasurface is reflectionless, we have that $\overline{\overline{S}}_{11} = \overline{\overline{S}}_{22} = 0$. Moreover, we also require it to be reciprocal, which leads to $\overline{\overline{S}}_{21} = \overline{\overline{S}}_{12}^T$. The matrix $\overline{\overline{S}}_{21}$ is straightforwardly defined from the specified $\pi/3$-polarization rotation angle as

$$\overline{\overline{S}}_{21} = \begin{pmatrix} \cos(\pi/3) & \sin(\pi/3) \\ -\sin(\pi/3) & \cos(\pi/3) \end{pmatrix} = \begin{pmatrix} 1/2 & \sqrt{3}/2 \\ -\sqrt{3}/2 & 1/2 \end{pmatrix}. \tag{4.45}$$

Upon insertion of these specifications of the scattering parameters into (4.13) and (4.14), the metasurface susceptibilities are obtained by the matrix inversion of (4.11). This leads to the nonzero susceptibilities

$$\chi_{em}^{xx} = \chi_{em}^{yy} = -\frac{2}{\sqrt{3}k_0}j, \tag{4.46a}$$

$$\chi_{me}^{xx} = \chi_{me}^{yy} = \frac{2}{\sqrt{3}k_0}j. \tag{4.46b}$$

These susceptibilities correspond to a chiral bianisotropic metasurface. As the previously synthesized gyrotropic metasurface, it performs the specified $\pi/3$-rotation of polarization irrespective of the incident wave polarization state. However, it is reciprocal, gainless, and lossless, as can be verified with (2.77). This is therefore the most practical one for reciprocal polarization rotation.

In this section, we have considered only four different sets of susceptibilities to perform the same transformation. In fact, there are many more possible susceptibility combinations that may be considered. However, most of them correspond to unpractical structures that may be lossy, active, nonreciprocal, and/or only perform the specified transformation for the specified polarization state of the incident wave.

4.1.7.2 Multiple Nonreciprocal Transformations

Consider a double transformation with a homoanisotropic metasurface whose susceptibilities are given in (4.10) in terms of the specified fields. The two transformations that are used to synthesize the metasurface are depicted in Figure 4.7. The first transformation, shown in Figure 4.7a, consists of a normally incident plane wave being fully reflected at a 45° angle. The second transformation, shown in Figure 4.7b, consists in the full absorption of an incident wave impinging on the metasurface at a 45° angle. In these two transformations, the transmitted field is set to zero and all the specified nonzero transverse components of the electric fields at $z = 0$ are

$$\mathbf{E}_{i,1} = \frac{\sqrt{2}}{2}(\hat{\mathbf{x}} + \hat{\mathbf{y}}), \tag{4.47a}$$

$$\mathbf{E}_{r,1} = \frac{\sqrt{2}}{2}(-\cos\theta_r\hat{\mathbf{x}} + \hat{\mathbf{y}})e^{-jk_x x}, \tag{4.47b}$$

$$\mathbf{E}_{i,2} = \frac{\sqrt{2}}{2}(\cos\theta_i\hat{\mathbf{x}} + \hat{\mathbf{y}})e^{-jk_x x}. \tag{4.47c}$$

The metasurface susceptibilities are found by substituting the electric fields given in (4.47), along with their magnetic counterparts, into (4.10). The resulting synthesized metasurface corresponds to a nonreciprocal structure, which exhibits complicated spatial variations of electric and magnetic loss and gain. For the sake of conciseness, we do not show the susceptibility functions here.

We now present full-wave simulations demonstrating the double transformation control capability of the synthesized metasurface. The simulations are performed in COMSOL, where the metasurface is implemented as a thin material slab of thickness $t = \lambda_0/100$ with "diluted" susceptibilities using (3.22), as explained in Section 3.2. The simulation corresponding to the transformation of Figure 4.7a is shown in Figure 4.8a, while the simulation corresponding to the transformation of Figure 4.7b is shown in Figure 4.8b. As can be seen, the metasurface scattering response seems to be in good agreement with the specifications, except for some undesired scattering due to imperfect slab modeling of the metasurface.

To verify the nonreciprocal nature of this metasurface, we excite it with an incident wave impinging on it at an angle $\theta_i = -45°$, which corresponds to the transformation reciprocal to that in Figure 4.7a. The corresponding simulation is shown in Figure 4.9. A comparison of the ratio of the power transfer between the normally incident wave and the obliquely reflected wave in Figure 4.8a and the power transfer between the obliquely incident wave and the normally reflected wave in Figure 4.9 clearly shows that the metasurface is nonreciprocal as it exhibits more dissipation in the simulation depicted in Figure 4.9 compared to that in Figure 4.8a.

4.1.7.3 Angle-Dependent Transformations

We now consider another metasurface synthesis example consisting in synthesizing a metasurface able to perform different transformations depending on the angle of incidence [39, 50, 134].

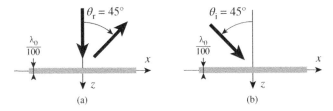

Figure 4.7 Illustration of the two specified transformations. (a) The normally incident plane wave is fully reflected at a 45° angle. (b) The obliquely incident plane wave is fully absorbed.

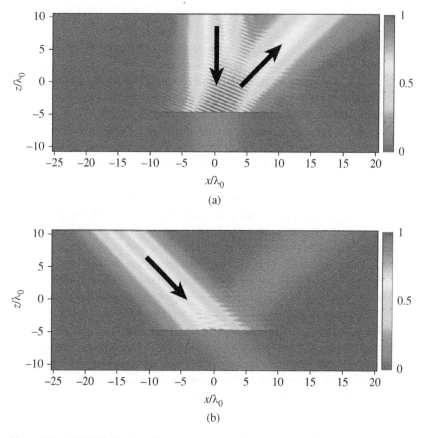

Figure 4.8 COMSOL simulated absolute value of the total electric field (V/m) corresponding to: (a) the transformation in Figure 4.7a, and (b) the transformation in Figure 4.7b.

For simplicity, we consider a uniform metasurface, which transforms the phase and the amplitude of the scattered waves without altering their direction of propagation. The metasurface is synthesized by specifying the reflection and transmission coefficients for three different incidence angles, which, by continuity, allows a relatively smooth control of the angular scattering as function of the incidence angle. The synthesis of a metasurface performing three transformations requires a number of degrees of freedom, which are here obtained by leveraging bianisotropy and making use of normal susceptibilities.

Let us consider the electromagnetic transformations depicted in Figure 4.10 where *p*-polarized incident plane waves are scattered, without deflection or polarization rotation, by a bianisotropic reflectionless metasurface.

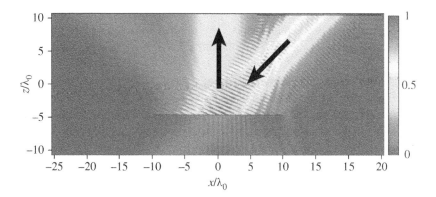

Figure 4.9 COMSOL simulated absolute value of the total electric field (V/m) corresponding to the reciprocal transformation of the one in Figure 4.7a.

Figure 4.10 Multiple scattering from a uniform reflectionless bianisotropic metasurface.

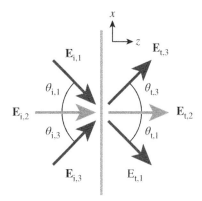

In such a transformation, the only nonzero electromagnetic field components are E_x, E_z, and H_y, and therefore only a few susceptibility components will be excited. Specifically, we find

$$\overline{\overline{\chi}}_{ee} = \begin{pmatrix} \chi_{ee}^{xx} & 0 & \chi_{ee}^{xz} \\ 0 & 0 & 0 \\ \chi_{ee}^{zx} & 0 & \chi_{ee}^{zz} \end{pmatrix}, \quad \overline{\overline{\chi}}_{em} = \begin{pmatrix} 0 & \chi_{em}^{xy} & 0 \\ 0 & 0 & 0 \\ 0 & \chi_{em}^{zy} & 0 \end{pmatrix}, \tag{4.48a}$$

$$\overline{\overline{\chi}}_{me} = \begin{pmatrix} 0 & 0 & 0 \\ \chi_{me}^{yx} & 0 & \chi_{me}^{yz} \\ 0 & 0 & 0 \end{pmatrix}, \quad \overline{\overline{\chi}}_{mm} = \begin{pmatrix} 0 & 0 & 0 \\ 0 & \chi_{mm}^{yy} & 0 \\ 0 & 0 & 0 \end{pmatrix}, \tag{4.48b}$$

where all the susceptibilities that are not excited by the specified fields have been set to zero for simplicity. The susceptibility tensors in (4.48) contain a total number of nine unknown components. However, we are here interested in synthesizing a *reciprocal* metasurface, which reduces the number of unknowns to six since the reciprocity conditions in (2.52) imply that $\chi_{ee}^{xz} = \chi_{ee}^{zx}$, $\chi_{em}^{xy} = -\chi_{me}^{yx}$, and $\chi_{em}^{zy} = -\chi_{me}^{yz}$.

Given the assumption of direction continuity, the metasurface is uniform in the xy-plane. Accordingly, the susceptibilities are independent of x and y, and hence the spatial derivatives on the right-hand sides of (4.1a) and (4.1b) only apply to the fields and not to the susceptibilities. This restriction means that the reflection and transmission angles follow conventional Snell's law, i.e. $\theta_r = -\theta_i$ and $\theta_t = \theta_i$.

Substituting the susceptibilities (4.48) into (4.1a) and (4.1b) with (3.47) and enforcing reciprocity leads to the two equations

$$\Delta H_y = -j\omega\epsilon_0(\chi_{ee}^{xx}E_{x,av} + \chi_{ee}^{xz}E_{z,av}) - jk_0\chi_{em}^{xy}H_{y,av}, \tag{4.49a}$$

$$\Delta E_x = -j\omega\mu_0\chi_{mm}^{yy}H_{y,av} + jk_0(\chi_{em}^{xy}E_{x,av} + \chi_{em}^{zy}E_{z,av})$$
$$- \chi_{ee}^{xz}\partial_x E_{x,av} - \chi_{ee}^{zz}\partial_x E_{z,av} - \eta_0\chi_{em}^{zy}\partial_x H_{y,av}. \tag{4.49b}$$

This system contains six unknown susceptibilities. In order to solve it, we apply the multiple transformation concept discussed in Section 4.1.3, which consists in specifying three independent sets of incident, reflected, and transmitted waves. Moreover, since the metasurface is perfectly uniform, we synthesize the metasurface using scattering parameters instead of fields, as done for the chiral metasurface in Section 4.1.7.1. Thus, the reflection (R) and transmission (T) coefficients of the metasurface in Figure 4.10 may be specified for three different angles of incidence. By specifying the reflection and transmission coefficients for three specific angles, one can achieve controllable transmission level quasi-continuous angular scattering since the response of the metasurface for non-specified angles de facto corresponds to an interpolation of the three specified responses.

To illustrate this concept, consider a reflectionless transformation ($R = 0$) where three incident plane waves, impinging on the metasurface at $\theta_{i,1} = -45°, \theta_{i,2} = 0°$, and $\theta_{i,3} = +45°$, are transmitted with transmission coefficients $T_1 = 0.75$, $T_2 = 0.5e^{j45°}$, and $T_3 = 0.25$ and transmission angles $\theta_t = \theta_i$. The susceptibilities corresponding to these specified scattering parameters are given by

$$\overline{\overline{\chi}}_{ee} = \frac{1}{k_0}\begin{pmatrix} -2.46 - 0.08j & 0 & 0.64j \\ 0 & 0 & 0 \\ 0.64j & 0 & 0.42 + 0.88j \end{pmatrix}, \quad \overline{\overline{\chi}}_{em} = \frac{1}{k_0}\begin{pmatrix} 0 & 1.74 - 0.68j & 0 \\ 0 & 0 & 0 \\ 0 & 0 & 0 \end{pmatrix},$$

$$\tag{4.50a}$$

$$\overline{\overline{\chi}}_{\text{mc}} = \frac{1}{k_0}\begin{pmatrix} 0 & 0 & 0 \\ -1.74 + 0.68j & 0 & 0 \\ 0 & 0 & 0 \end{pmatrix}, \quad \overline{\overline{\chi}}_{\text{mm}} = \frac{1}{k_0}\begin{pmatrix} 0 & 0 & 0 \\ 0 & 1.02 - 1.45j & 0 \\ 0 & 0 & 0 \end{pmatrix}.$$

(4.50b)

The expressions relating the susceptibilities to the reflection and transmission coefficients, being excessively long, are not provided here. It can be easily verified that these susceptibility tensors satisfy the reciprocity conditions (2.52), while the complex values of the susceptibilities indicate the presence of both gain and loss.

To verify that the scattered waves have the specified amplitude and phase at the three specified incidence angles and also to examine the response at non-specified angles, we analyze the scattering from the synthesized metasurface versus the incidence angle. For this purpose, we solve the relations (4.49) to determine the reflection and transmission coefficients versus θ_i. The resulting amplitude and phase of the reflection and transmission coefficients are plotted in Figure 4.11a and b, respectively. The metasurface is seen to exhibit the specified response in terms of both coefficients at the three specified angles. Moreover, the transmission exhibits a continuous amplitude decrease as θ_i increases beyond $-50°$.

We shall now apply the same synthesis procedure to another interesting example: a spatial angular phaser, specified to be reflectionless, gainless, and lossless and exhibiting a transmission phase shift, ϕ, that is function of the incidence angle, i.e. $T = e^{j\phi(\theta_i)}$. The three incident plane waves impinge on the metasurface at $\theta_{i,1} = -45°$, $\theta_{i,2} = 0°$, and $\theta_{i,3} = +45°$ and are transmitted with

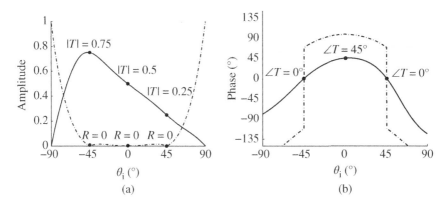

Figure 4.11 Reflection (dashed line) and transmission (solid line) amplitude (a) and phase (b) as functions of the incidence angle for a metasurface synthesized for the transmission coefficients $T = \{0.75; 0.5e^{j45°}; 0.25\}$ (and $R = 0$) at the respective incidence angles $\theta_i = \{-45°; 0°; +45°\}$.

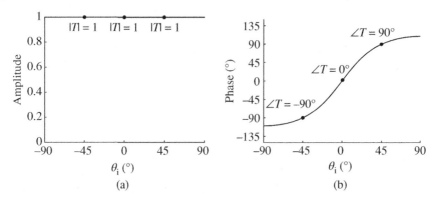

Figure 4.12 Transmission amplitude (a) and phase (b) as functions of the incidence angle for a metasurface synthesized for the transmission coefficients $T = \{e^{-j90°}; 1; e^{j90°}\}$ (and $R = 0$) at the respective incidence angles $\theta_i = \{-45°; 0°; +45°\}$.

transmission coefficients $T_1 = e^{-j\alpha}$, $T_2 = 1$, and $T_3 = e^{j\alpha}$, where α is a specified phase shift. Solving relations (4.49) for these specifications yields the nonzero susceptibilities

$$\chi_{ee}^{xz} = \chi_{ee}^{zx} = \frac{2\sqrt{2}}{k_0} \tan\left(\frac{\alpha}{2}\right). \tag{4.51}$$

Being restricted to these two susceptibilities, this metasurface exhibits the simple angle-dependent transmission coefficient

$$T(\theta_i) = -1 + \frac{2}{1 - j\sqrt{2}\sin(\theta_i)\tan\left(\frac{\alpha}{2}\right)}, \tag{4.52}$$

while the reflection coefficient is $R(\theta_i) = 0$.

Figure 4.12 plots the angular response of the transmission coefficient in (4.52) for a specified phase shift of $\alpha = 90°$. As expected, the transmission amplitude remains unity for all incidence angles while the transmission phase is asymmetric around broadside and covers about a 220°-phase range. It is interesting to see that such kind of angular asymmetric response is due to the susceptibilities χ_{ee}^{xz} and χ_{ee}^{zx}, as we shall again discuss in Section 6.4.1.

4.2 Time-Varying Metasurfaces

4.2.1 Formulation of the Problem

A time-varying metasurface is a metasurface whose medium parameters, i.e. susceptibilities, vary in time. A time-varying metasurface enables useful phenomena that do not exist in conventional time-invariant metasurfaces, such as frequency generation and mixing, attenuation, or amplification [24]. This type of metasurface must be logically modeled in the time domain to properly account for the time-varying nature of its susceptibilities. Therefore, we next make use of the time-domain expressions of the GSTCs (3.46), which we repeat here for convenience

$$\hat{\mathbf{r}} \times \Delta \mathcal{H}(t) = \frac{\partial}{\partial t} \mathbf{P}_{\parallel}(t) - \hat{\mathbf{r}} \times \nabla_{\parallel} \mathcal{M}_r(t), \tag{4.53a}$$

$$\hat{\mathbf{r}} \times \Delta \mathcal{E}(t) = -\mu_0 \frac{\partial}{\partial t} \mathcal{M}_{\parallel}(t) - \hat{\mathbf{r}} \times \nabla_{\parallel}(\mathcal{P}_r(t)/\epsilon_0). \tag{4.53b}$$

In general, the susceptibilities, in addition to being time-variant, are also temporally dispersive. This implies that the polarization densities in Eq. (4.53) must be proportional to the time convolution between the fields and the susceptibilities, as mentioned in Section 2.2.

It is obviously much more difficult to synthesize a time-varying metasurface than a time-invariant one, even if we are only interested in the mathematical synthesis and without concern for its fabrication. For this reason, we shall next restrict our attention to nondispersive metasurfaces.[4] This is justified by the fact that the majority of current time-varying metamaterial structures are based on non-resonant systems and thus exhibit weak (negligible) temporal dispersion. This is for instance the case of space–time modulated leaky-way system presented in [163]. Such systems are typically made of varactor-loaded transmission lines, whose varactors are temporally modulated so as to vary the characteristic impedance of the transmission line. This temporal modulation may be slower or faster than the frequency of the signal propagating on the transmission line and lead to various interesting electromagnetic effects [24].

The advantage of nondispersive medium modeling is that one can ignore the time convolutions in Eq. (4.53). It follows that the polarization densities may be simply expressed as the multiplication of the susceptibilities and fields. If we also assume the absence of normal polarizations, the time-domain GSTCs in (4.53), with the bianisotropic definition of the polarization densities, reduce to

$$\hat{\mathbf{r}} \times \Delta \mathcal{H}(t) = \epsilon_0 \frac{\partial}{\partial t} \left[\overline{\overline{\chi}}_{\mathrm{ee}}(t) \cdot \mathcal{E}_{\mathrm{av}}(t) + \overline{\overline{\chi}}_{\mathrm{em}}(t) \cdot \mathcal{H}_{\mathrm{av}}(t) \right], \tag{4.54a}$$

$$\hat{\mathbf{r}} \times \Delta \mathcal{E}(t) = -\mu_0 \frac{\partial}{\partial t} \left[\overline{\overline{\chi}}_{\mathrm{mm}}(t) \cdot \mathcal{H}_{\mathrm{av}}(t) + \overline{\overline{\chi}}_{\mathrm{me}}(t) \cdot \mathcal{E}_{\mathrm{av}}(t) \right]. \tag{4.54b}$$

Equations (4.54) form a set of coupled differential equations, whose resolution for the time-domain susceptibilities is much more difficult than that of the time-invariant metasurfaces discussed in Section 4.1. In what follows, we provide a simple example in order to illustrate the procedure.

---

4 Note that we provide a synthesis example of a dispersive time-varying metasurface in Section 5.3.2.

4.2.2 Harmonic-Generation Time-Varying Metasurface

Let us consider a homoisotropic ($\overline{\overline{\chi}}_{em} = \overline{\overline{\chi}}_{me} = 0$) reflectionless metasurface lying in the xy-plane that transforms a normally incident plane wave, with frequency ω_1, into a normally transmitted plane wave, with frequency ω_2. Since the metasurface is homoisotropic, Eq. (4.54) reduces to

$$\hat{z} \times \Delta \mathcal{H}(t) = \epsilon_0 \left[\mathcal{E}_{av}(t) \cdot \frac{\partial}{\partial t} \overline{\overline{\chi}}_{ee}(t) + \overline{\overline{\chi}}_{ee}(t) \cdot \frac{\partial}{\partial t} \mathcal{E}_{av}(t) \right], \tag{4.55a}$$

$$\hat{z} \times \Delta \mathcal{E}(t) = -\mu_0 \left[\mathcal{H}_{av}(t) \cdot \frac{\partial}{\partial t} \overline{\overline{\chi}}_{mm}(t) + \overline{\overline{\chi}}_{mm}(t) \cdot \frac{\partial}{\partial t} \mathcal{H}_{av}(t) \right], \tag{4.55b}$$

where we have used the product rule in expanding the time derivatives. We next consider the case of x-polarized plane waves and assume no polarization rotation. This reduces (4.55) further to

$$\Delta \mathcal{H}_y(t) = -\epsilon_0 \left[\mathcal{E}_{x,av}(t) \cdot \frac{\partial}{\partial t} \tilde{\chi}_{ee}^{xx}(t) + \tilde{\chi}_{ee}^{xx}(t) \cdot \frac{\partial}{\partial t} \mathcal{E}_{x,av}(t) \right], \tag{4.56a}$$

$$\Delta \mathcal{E}_x(t) = -\mu_0 \left[\mathcal{H}_{y,av}(t) \cdot \frac{\partial}{\partial t} \tilde{\chi}_{mm}^{yy}(t) + \tilde{\chi}_{mm}^{yy}(t) \cdot \frac{\partial}{\partial t} \mathcal{H}_{y,av}(t) \right], \tag{4.56b}$$

These two equations may be individually solved for the time-domain suscepti-bilities. For that purpose, we consider the following instantaneous electric and magnetic fields for the incident and transmitted waves at $z = 0$:

$$\mathcal{E}_{i,x} = \eta_0 \mathcal{H}_{i,y} = A_i \cos(\omega_1 t) \quad \text{and} \quad \mathcal{E}_{t,x} = \eta_0 \mathcal{H}_{t,y} = A_t \cos(\omega_2 t), \tag{4.57}$$

corresponding to the frequency transformation of plane waves propagating in vac-uum with field amplitudes A_i and A_t. Substituting (4.57) into (4.56) and solving these two differential equations yields the closed-form susceptibilities

$$\tilde{\chi}_{ee}(t) = \frac{2 A_i \omega_2 \sin \left(\omega_1 t \right) + \omega_1 \left[C_e \eta_0 \omega_2 \epsilon_0 - 2 A_t \sin \left(\omega_2 t \right) \right]}{A_i \eta_0 \omega_1 \omega_2 \epsilon_0 \cos \left(\omega_1 t \right) + A_t \eta_0 \omega_1 \omega_2 \epsilon_0 \cos \left(\omega_2 t \right)}, \tag{4.58a}$$

$$\tilde{\chi}_{mm}(t) = \frac{2 A_i \eta_0 \omega_2 \sin \left(\omega_1 t \right) + \omega_1 \left[C_m \mu_0 \omega_2 - 2 A_t \eta_0 \sin \left(\omega_2 t \right) \right]}{\mu_0 \omega_1 \omega_2 \left[A_i \cos \left(\omega_1 t \right) + At \cos \left(\omega_2 t \right) \right]}, \tag{4.58b}$$

where C_e and C_m are integration constants.

We may simplify this result by considering the particular case of second-harmonic generation, i.e. $\omega_2 = 2\omega_1$. If we also assume that $C_e = C_m = 0$, we obtain

$$\tilde{\chi}_{ee}(t) = \tilde{\chi}_{mm}(t) = \frac{2 \left[A_i - A_t \cos \left(\omega_1 t \right) \right] \sin \left(\omega_1 t \right)}{k_1 \left[A_i \cos \left(\omega_1 t \right) + A_t \cos \left(2\omega_1 t \right) \right]}, \tag{4.59}$$

where $k_1 = \omega_1/c_0$. Note that the first equality is the reflectionless condition, which we established in Section 4.1.4.

This simple example illustrates how the GSTCs may be solved in the time domain to synthesize a time-varying metasurface. The synthesis procedure is identical to that for time-invariant metasurfaces except that it is necessary to solve a time-domain differential set of equations, which may be generally not trivial to achieve.

4.3 Nonlinear Metasurfaces

The concept of nonlinear metamaterials and, more specifically, nonlinear meta-surfaces has been considered a paradigm shift in nonlinear optics [8, 83, 105]. One reason is the high near-field enhancement achievable with metasurfaces, leading to extreme nonlinear responses. Another reason is the unprecedented capabilities of metasurfaces to control nonlinear responses via complex scattering particles geometries, an impossible feat with conventional nonlinear crystals. Nonlinear metasurfaces have thus opened up a new research avenue in the field of nonlinear optics, with promises for new applications [8, 83, 105].

As we shall see next, a nonlinear metasurface exhibits many more degrees free-dom, in the form of nonlinear susceptibility tensors, than a linear metasurface. This requires an adequate modeling approach that takes into account all the pos-sible interactions of the fields with the metasurface. Fortunately, the GSTCs per-fectly fill that need.

For convenience, we will focus our attention on the nonlinear phenomenon of second-harmonic generation (SHG), which is a second-order nonlinear process [21]. The formalism that we will present can then be easily extended to higher order nonlinear processes by including higher order nonlinear susceptibility tensors.

4.3.1 Second-Order Nonlinearity

A second-order nonlinear metasurface is characterized by the presence of (first-order) linear and (second-order) nonlinear susceptibility tensors. When such a metasurface is illuminated by a wave of frequency ω, the interactions between the wave and the nonlinear susceptibilities essentially result in scattered waves of frequencies ω and 2ω. Note that the superposition of the scattered waves and the incident wave may also result in the generation of higher order harmonic, but those are generally much weaker than the first- and second-order ones and are therefore usually neglected [21].

We will now present two approaches for modeling second-order nonlinear meta-surfaces. Both are based on the GSTCs, one in the frequency and one in the time domain.

4.3.1.1 Frequency-Domain Approach

The frequency-domain model is based on the following two main assump-tions: (i) the waves interacting with the metasurface are *only* of frequencies ω

and 2ω, and (ii) the incident wave (the pump) is not depleted, i.e. the power transfer between the pump and the nonlinear scattered light is negligible. The latter assumption violates energy conservation but is justified by the fact that the amplitude of the nonlinear scattering is several orders of magnitude smaller than that of the pump [21].

Based on these assumptions, we can independently model the interactions at ω and those at 2ω in the GSTCs. This leads to the following GSTC relations, relating the fields and polarizations at 2ω, for a metasurface lying in the xy-plane:

$$\mathbf{z} \times \Delta\mathbf{H}^{2\omega} = j2\omega\mathbf{P}^{2\omega} - \mathbf{z} \times \nabla M_z^{2\omega}, \tag{4.60a}$$

$$\mathbf{z} \times \Delta\mathbf{E}^{2\omega} = -j2\omega\mu_0\mathbf{M}^{2\omega} - \frac{1}{\epsilon_0}\mathbf{z} \times \nabla P_z^{2\omega}, \tag{4.60b}$$

where the polarization densities are given as the sum of linear and nonlinear terms as

$$\mathbf{P}^{2\omega} = \mathbf{P}_{lin}^{2\omega} + \mathbf{P}_{nl}^{2\omega}, \tag{4.61a}$$

$$\mathbf{M}^{2\omega} = \mathbf{M}_{lin}^{2\omega} + \mathbf{M}_{nl}^{2\omega}. \tag{4.61b}$$

The linear polarization densities correspond to the interactions of the fields with the metasurface at frequency 2ω. Therefore, the corresponding susceptibilities must also be evaluated at frequency 2ω. This leads to the following expressions for the linear polarization densities based on Eq. (2.2):

$$\mathbf{P}_{lin}^{2\omega} = \epsilon_0\overline{\overline{\chi}}_{ee}(2\omega) \cdot \mathbf{E}_{av}^{2\omega} + \epsilon_0\eta_0\overline{\overline{\chi}}_{em}(2\omega) \cdot \mathbf{H}_{av}^{2\omega}, \tag{4.62a}$$

$$\mathbf{M}_{lin}^{2\omega} = \overline{\overline{\chi}}_{mm}(2\omega) \cdot \mathbf{H}_{av}^{2\omega} + \frac{1}{\eta_0}\overline{\overline{\chi}}_{me}(2\omega) \cdot \mathbf{E}_{av}^{2\omega}. \tag{4.62b}$$

On the other hand, the nonlinear polarization densities correspond to the interactions of the fields at the frequency ω with the metasurface with second-order susceptibilities also evaluated at ω. These nonlinear polarizations thus play the role of nonlinear sources from which the second-harmonic wave is generated. To take into account all possible interactions, the nonlinear polarization densities are expressed as

$$\mathbf{P}_{nl}^{2\omega} = \frac{1}{2}\epsilon_0 \left(\overline{\overline{\chi}}_{eee}(\omega) : \mathbf{E}_{av}^{\omega}\mathbf{E}_{av}^{\omega} + \eta_0\overline{\overline{\chi}}_{eem}(\omega) : \mathbf{E}_{av}^{\omega}\mathbf{H}_{av}^{\omega} \right.$$
$$\left. + \eta_0^2\overline{\overline{\chi}}_{emm}(\omega) : \mathbf{H}_{av}^{\omega}\mathbf{H}_{av}^{\omega} \right), \tag{4.63a}$$

$$\mathbf{M}_{nl}^{2\omega} = \frac{1}{2} \left(\eta_0\overline{\overline{\chi}}_{mmm}(\omega) : \mathbf{H}_{av}^{\omega}\mathbf{H}_{av}^{\omega} + \overline{\overline{\chi}}_{mem}(\omega) : \mathbf{E}_{av}^{\omega}\mathbf{H}_{av}^{\omega} \right.$$
$$\left. + \frac{1}{\eta_0}\overline{\overline{\chi}}_{mee}(\omega) : \mathbf{E}_{av}^{\omega}\mathbf{E}_{av}^{\omega} \right). \tag{4.63b}$$

In these relations, we have used our own convention that the nonlinear susceptibilities all have the same dimension, i.e. they are all expressed in (m^2/V), while the linear ones are expressed in (m). We have also grouped the susceptibility tensors $\overline{\overline{\chi}}_{\text{eem}}$ and $\overline{\overline{\chi}}_{\text{eme}}$, and $\overline{\overline{\chi}}_{\text{mem}}$ and $\overline{\overline{\chi}}_{\text{mme}}$, since their effects are indistinguishable from each other.

Note that we have introduced a factor $\frac{1}{2}$ in Eq. (4.63) that is not present in Eq. (4.62). This is because the time derivatives of the fields, which must be evaluated to provide the nonlinear GSTCs in Eq. (4.60) from the general time-domain GSTCs in Eq. (3.46), yield the different results

$$\frac{\partial}{\partial t} \mathbf{E}_{\text{av}}^{2\omega} \propto \frac{\partial}{\partial t} \cos(2\omega t) = -2\omega \sin(2\omega t), \tag{4.64}$$

and

$$\frac{\partial}{\partial t} \mathbf{E}_{\text{av}}^{\omega} \mathbf{E}_{\text{av}}^{\omega} \propto \frac{\partial}{\partial t} \cos^2(\omega t) = -\omega \sin(2\omega t). \tag{4.65}$$

The general nonlinear second-order GSTCs are now readily obtained by substituting Eqs. (4.62) and (4.63) into Eq. (4.60). One may then either synthesize or analyze the scattering from a nonlinear metasurface following the procedures described for linear metasurfaces in Section 4.1. We shall now derive the expressions relating the second-harmonic fields scattered by *uniform* nonlinear metasurface illuminated by a normally incident plane wave to its linear and nonlinear susceptibilities. These expressions may then be used either to find the required susceptibilities so as to achieve specified linear and nonlinear scattering characteristics or, alternatively, to obtain the fields scattered by a metasurface with known susceptibilities.

To derive these expressions, we start by writing the magnetic fields of the forward (fw) propagating waves (+z) and the backward (bw) propagating waves (−z). Since we are considering only normally propagating plane waves, we simply have

$$\mathbf{H}_{\text{fw}} = \frac{1}{\eta_0} \overline{\overline{J}} \cdot \mathbf{E}_{\text{fw}}, \quad \mathbf{H}_{\text{bw}} = -\frac{1}{\eta_0} \overline{\overline{J}} \cdot \mathbf{E}_{\text{bw}}, \tag{4.66}$$

where $\overline{\overline{J}}$ is the 90°-rotation matrix defined as

$$\overline{\overline{J}} = \begin{pmatrix} 0 & -1 \\ 1 & 0 \end{pmatrix}. \tag{4.67}$$

Then, we define the averages of the linear fields in Eq. (4.63). For this purpose, we make use of Eq. (4.66) and of the scattering matrices, which provide the complex amplitudes of the reflected and transmitted waves, that were derived in Section 4.1.4. In the case of an incident wave propagating in the forward direction, we obtain

$$\mathbf{E}_{\text{av,fw}}^{\omega} = \frac{1}{2} \left(\overline{\overline{I}} + \overline{\overline{S}}_{11} + \overline{\overline{S}}_{21} \right) \cdot \mathbf{E}_0^{\omega}, \tag{4.68a}$$

$$\mathbf{H}_{\text{av,fw}}^{\omega} = \frac{1}{2\eta_0} \overline{\overline{J}} \cdot \left(\overline{\overline{I}} - \overline{\overline{S}}_{11} + \overline{\overline{S}}_{21} \right) \cdot \mathbf{E}_0^{\omega}, \tag{4.68b}$$

where $\mathbf{E}_0^\omega$ is the amplitude of the pump electric field and $\bar{\bar{I}}$ is the two-dimensional identity matrix. In these expressions, port 1 and port 2 refer to the left ($z < 0$) and right ($z > 0$) sides of the metasurface, respectively. Now, in the case of backward propagating incident wave, we get

$$\mathbf{E}_{av,bw}^\omega = \frac{1}{2} \left(\bar{\bar{I}} + \bar{\bar{S}}_{22} + \bar{\bar{S}}_{12} \right) \cdot \mathbf{E}_0^\omega, \tag{4.69a}$$

$$\mathbf{H}_{av,bw}^\omega = -\frac{1}{2\eta_0}\bar{\bar{J}} \cdot \left(\bar{\bar{I}} - \bar{\bar{S}}_{22} + \bar{\bar{S}}_{12} \right) \cdot \mathbf{E}_0^\omega. \tag{4.69b}$$

Applying the same procedure, we can find the averages and differences of the nonlinear fields at frequency 2ω that are required in Eq. (4.62). Since there is no incident nonlinear wave, the former is simply given by

$$\mathbf{E}_{av}^{2\omega} = \frac{1}{2} \left(\mathbf{E}_{fw}^{2\omega} + \mathbf{E}_{bw}^{2\omega} \right), \tag{4.70a}$$

$$\mathbf{H}_{av}^{2\omega} = \frac{1}{2\eta_0}\bar{\bar{J}} \cdot \left(\mathbf{E}_{fw}^{2\omega} - \mathbf{E}_{bw}^{2\omega} \right), \tag{4.70b}$$

while the latter reads

$$\Delta\mathbf{E}^{2\omega} = \mathbf{E}_{fw}^{2\omega} - \mathbf{E}_{bw}^{2\omega}, \tag{4.71a}$$

$$\Delta\mathbf{H}^{2\omega} = \frac{1}{\eta_0}\bar{\bar{J}} \cdot \left(\mathbf{E}_{fw}^{2\omega} + \mathbf{E}_{bw}^{2\omega} \right). \tag{4.71b}$$

Substituting (4.70) and (4.71), along with (4.61) and (4.62), into (4.60), reduces the nonlinear GSTCs to

$$\mathbf{E}_{fw}^{2\omega} + \mathbf{E}_{bw}^{2\omega} = -j\frac{\omega}{c_0} \left[\bar{\bar{\chi}}_{ee} \cdot \left(\mathbf{E}_{fw}^{2\omega} + \mathbf{E}_{bw}^{2\omega} \right) \right.$$
$$\left. + \bar{\bar{\chi}}_{em} \cdot \bar{\bar{J}} \cdot \left(\mathbf{E}_{fw}^{2\omega} - \mathbf{E}_{bw}^{2\omega} \right) \right] - j\omega\eta_0 \mathbf{P}_{nl}^{2\omega}, \tag{4.72a}$$

$$\bar{\bar{J}} \cdot \left(\mathbf{E}_{fw}^{2\omega} - \mathbf{E}_{bw}^{2\omega} \right) = -j\frac{\omega}{c_0} \left[\bar{\bar{\chi}}_{mm} \cdot \bar{\bar{J}} \cdot \left(\mathbf{E}_{fw}^{2\omega} - \mathbf{E}_{bw}^{2\omega} \right) \right.$$
$$\left. + \bar{\bar{\chi}}_{me} \cdot \left(\mathbf{E}_{fw}^{2\omega} + \mathbf{E}_{bw}^{2\omega} \right) \right] - j\omega\mu_0 \mathbf{M}_{nl}^{2\omega}, \tag{4.72b}$$

where c_0 is the speed of light in vacuum.

In order to express the forward and backward nonlinear fields scattered by the metasurface in terms of its susceptibilities, we may now solve Eq. (4.72), which yields

$$\mathbf{E}_{bw}^{2\omega} = \frac{1}{\epsilon_0}\bar{\bar{C}}_1 \cdot \mathbf{P}_{nl}^{2\omega} + \eta_0 \bar{\bar{C}}_2 \cdot \bar{\bar{J}} \cdot \mathbf{M}_{nl}^{2\omega}, \tag{4.73a}$$

$$\mathbf{E}_{fw}^{2\omega} = \frac{1}{\epsilon_0}\bar{\bar{C}}_3 \cdot \mathbf{P}_{nl}^{2\omega} - \eta_0 \bar{\bar{C}}_4 \cdot \bar{\bar{J}} \cdot \mathbf{M}_{nl}^{2\omega}, \tag{4.73b}$$

where the matrices $\bar{\bar{C}}$ are given by

$$\bar{\bar{C}}_1 = -j2\bar{\bar{A}} \cdot \left(\frac{c_0}{\omega}\bar{\bar{J}} + j\bar{\bar{\chi}}_{mm} \cdot \bar{\bar{J}} + j\bar{\bar{\chi}}_{me} \right) \cdot \left(\frac{c_0}{\omega}\bar{\bar{I}} + j\bar{\bar{\chi}}_{ee} + j\bar{\bar{\chi}}_{em} \cdot \bar{\bar{J}} \right)^{-1}, \tag{4.74a}$$

$$\overline{\overline{C}}_2 = -j2\overline{\overline{A}} \cdot \overline{\overline{J}}, \tag{4.74b}$$

$$\overline{\overline{C}}_3 = j2\left(\frac{c_0}{\omega}\overline{\overline{I}} + j\overline{\overline{\chi}}_{ee} + j\overline{\overline{\chi}}_{em} \cdot \overline{\overline{J}}\right)^{-1} \cdot \left[\left(\frac{c_0}{\omega}\overline{\overline{I}} + j\overline{\overline{\chi}}_{ee} - j\overline{\overline{\chi}}_{em} \cdot \overline{\overline{J}}\right) \cdot \overline{\overline{A}}\right.$$
$$\left. \cdot \left(\frac{c_0}{\omega}\overline{\overline{J}} + j\overline{\overline{\chi}}_{mm} \cdot \overline{\overline{J}} + j\overline{\overline{\chi}}_{me}\right) \cdot \left(\frac{c_0}{\omega}\overline{\overline{I}} + j\overline{\overline{\chi}}_{ee} + j\overline{\overline{\chi}}_{em} \cdot \overline{\overline{J}}\right)^{-1} - \overline{\overline{I}}\right], \tag{4.74c}$$

$$\overline{\overline{C}}_4 = -j2\left(\frac{c_0}{\omega}\overline{\overline{I}} + j\overline{\overline{\chi}}_{ee} + j\overline{\overline{\chi}}_{em} \cdot \overline{\overline{J}}\right)^{-1} \cdot \left(\frac{c_0}{\omega}\overline{\overline{I}} + j\overline{\overline{\chi}}_{ee} - j\overline{\overline{\chi}}_{em} \cdot \overline{\overline{J}}\right) \cdot \overline{\overline{A}} \cdot \overline{\overline{J}}, \tag{4.74d}$$

$$\overline{\overline{A}} = \left[\left(\frac{c_0}{\omega}\overline{\overline{J}} + j\overline{\overline{\chi}}_{mm} \cdot \overline{\overline{J}} - j\overline{\overline{\chi}}_{me}\right) + \left(\frac{c_0}{\omega}\overline{\overline{J}} + j\overline{\overline{\chi}}_{mm} \cdot \overline{\overline{J}} + j\overline{\overline{\chi}}_{me}\right)\right.$$
$$\left. \times \left(\frac{c_0}{\omega}\overline{\overline{I}} + j\overline{\overline{\chi}}_{ee} + j\overline{\overline{\chi}}_{em} \cdot \overline{\overline{J}}\right)^{-1} \cdot \left(\frac{c_0}{\omega}\overline{\overline{I}} + j\overline{\overline{\chi}}_{ee} - j\overline{\overline{\chi}}_{em} \cdot \overline{\overline{J}}\right)\right]^{-1}. \tag{4.74e}$$

Assuming that the metasurface susceptibilities are known, the procedure to obtain the nonlinear scattered fields is as follows: (i) use the procedure outlined in Section 4.1.4 to obtain the metasurface scattering matrices, (ii) use Eq. (4.68) or Eq. (4.69) depending on whether the incident propagates forward or backward to define the average linear fields, (iii) the forward/backward nonlinear fields may now be obtained from Eq. (4.73) by defining the $\overline{C}$ matrices in Eq. (4.74) and the nonlinear polarizations in Eq. (4.63).

Alternatively, the nonlinear susceptibilities of a metasurface may be obtained by performing linear and nonlinear periodic simulations and solving Eq. (4.73) for the susceptibilities.

4.3.1.2 Time-Domain Approach

We have just seen how the nonlinear fields scattered by a second-order nonlinear metasurface may be obtained using the frequency-domain GSTCs. The method presented in the previous section is relatively straightforward because it ignores the depletion of the pump (excitation) and also neglects the interactions between the incident wave and the nonlinear scattered wave. Finally, it only considers waves at frequencies ω and 2ω, and treats the linear and nonlinear scattering independently.

The time-domain method that we will describe now is more general than the frequency-domain one because it automatically takes into account the depletion of the pump and the interactions between waves at different frequencies [8]. As a consequence, it is also much more difficult to solve since it implies dealing with time-domain derivatives. We shall illustrate this time-domain method with a simplified example for simplicity and conciseness. More general cases may then be easily developed by extending the procedure.

Let us consider a second-order nonlinear *uniform* metasurface composed of only the linear susceptibilities χ_{ee}^{xx} and χ_{mm}^{yy}, and the nonlinear susceptibilities χ_{eee}^{xxx} and χ_{mmm}^{yyy}. For the sake of simplicity, we assume that the susceptibilities are dispersionless so as to avoid convolution products between fields and susceptibilities. Note that due to the nonlinear nature of the process, the fields scattered by the metasurface may typically be expressed as a sum of harmonics, i.e. $\mathbf{E}(\mathbf{r}, \omega) = \sum_{n=1}^{\infty} \mathbf{E}_n(\mathbf{r}) e^{jn\omega t}$, where the amplitude $\mathbf{E}_n$ of the harmonics decreases as n increases. This means that the dispersionless susceptibility condition may be essentially reduced to $\chi(\omega) = \chi(2\omega)$. From the time-domain GSTCs in Eq. (3.46), we have thus[5]

$$-\Delta\mathcal{H} = \epsilon_0 \chi_{ee}^{xx} \frac{\partial}{\partial t} \mathcal{E}_{av} + \epsilon_0 \chi_{eee}^{xxx} \frac{\partial}{\partial t} \mathcal{E}_{av}^2, \tag{4.75a}$$

$$-\Delta\mathcal{E} = \mu_0 \chi_{mm}^{yy} \frac{\partial}{\partial t} \mathcal{H}_{av} + \mu_0 \eta_0 \chi_{mmm}^{yyy} \frac{\partial}{\partial t} \mathcal{H}_{av}^2, \tag{4.75b}$$

where we consider that all the waves interacting with the metasurface are x-polarized. Since the metasurface is uniform, it scatters waves only normally when illuminated by a normally incident pump. Considering a forward propagating pump along with (4.66), Eq. (4.75) becomes

$$\begin{aligned}
-\mathcal{E}_{fw} - \mathcal{E}_{bw} + \mathcal{E}_{pump} &= \frac{\eta_0 \epsilon_0}{2} \chi_{ee}^{xx} \frac{\partial}{\partial t} \left(\mathcal{E}_{fw} + \mathcal{E}_{bw} + \mathcal{E}_{pump} \right) \\
&+ \frac{\eta_0 \epsilon_0}{4} \chi_{eee}^{xxx} \frac{\partial}{\partial t} \left(\mathcal{E}_{fw} + \mathcal{E}_{bw} + \mathcal{E}_{pump} \right)^2,
\end{aligned} \tag{4.76a}$$

$$\begin{aligned}
-\mathcal{E}_{fw} + \mathcal{E}_{bw} + \mathcal{E}_{pump} &= \frac{\mu_0}{2\eta_0} \chi_{mm}^{yy} \frac{\partial}{\partial t} \left(\mathcal{E}_{fw} - \mathcal{E}_{bw} + \mathcal{E}_{pump} \right) \\
&+ \frac{\mu_0}{2\eta_0} \chi_{mmm}^{yyy} \frac{\partial}{\partial t} \left(\mathcal{E}_{fw} - \mathcal{E}_{bw} + \mathcal{E}_{pump} \right)^2.
\end{aligned} \tag{4.76b}$$

From this system of equations, it is again possible to either find the susceptibilities, assuming that the time-domain values of the fields are known, or to compute the nonlinear scattered fields from the metasurface. The former operation is rather trivial since it reduces to solving a linear system of equations, while the latter is more complicated as it consists in solving a differential system of coupled nonlinear equations. We will next consider the latter operation.

Solving Eq. (4.76) analytically is unfortunately impossible. However, we may find an approximate expression for the scattered fields using perturbation theory. For this purpose, we assume that the scattered fields may be expanded as

$$\mathcal{E}_{fw} \approx \mathcal{E}_{0,fw} + \gamma \mathcal{E}_{1,fw} + \gamma^2 \mathcal{E}_{2,fw} + \cdots + \gamma^n \mathcal{E}_{n,fw}, \tag{4.77a}$$

$$\mathcal{E}_{bw} \approx \mathcal{E}_{0,bw} + \gamma \mathcal{E}_{1,bw} + \gamma^2 \mathcal{E}_{2,bw} + \cdots + \gamma^n \mathcal{E}_{n,bw}, \tag{4.77b}$$

where γ is a small quantity, typically of the same order of magnitude as the nonlinear susceptibilities, such that

$$\chi_{ee}^{xx} \gg \chi_{eee}^{xxx} \sim \gamma, \quad \text{and} \quad \chi_{mm}^{yy} \gg \chi_{mmm}^{yyy} \sim \gamma. \tag{4.78}$$

---

5 Note that the susceptibilities here are assumed to be time-independent so that the time derivatives only apply to the fields.

This assumption is justified by the fact that conventional nonlinear materials present a very small nonlinear response, which is typically in the order of $\gamma \sim 10^{-12}$ [21].

We may now solve Eq. (4.76) for the nonlinear scattered fields by inserting (4.77) and (4.78) into (4.76) and using $\mathcal{E}_{\text{pump}} = E_0 \cos(\omega t)$. We may remove all terms that are proportional to γ^m, where $m > n$ and thus solve for each value of n independently. We will next restrict our attention to the first four orders, i.e. $n = 0, 1, 2$, and 3. Note that the different value of n may be associated with different harmonics as

$$n = 0 \rightarrow \omega, \tag{4.79a}$$

$$n = 1 \rightarrow 2\omega, \tag{4.79b}$$

$$n = 2 \rightarrow \omega, 3\omega, \tag{4.79c}$$

$$n = 3 \rightarrow 2\omega, 4\omega. \tag{4.79d}$$

This means that to obtain the fields scattered by the metasurface at frequency ω, we must combine the contributions from the even values of n. On the other hand, the fields scattered at frequency 2ω are calculated from the odd values of n. By recursively solving Eq. (4.76) for each value of n up to $n = 3$, we obtain the following expressions for the fields scattered at frequency ω:

$$
\begin{aligned}
E_{\text{fw}}^{\omega} = E_0 & \frac{4 + \chi_{\text{ee}}^{xx} \chi_{\text{mm}}^{yy} k^2}{\left(2 + jk\chi_{\text{ee}}^{xx}\right)\left(2 + jk\chi_{\text{mm}}^{yy}\right)} \\
& - 4E_0^3 k^2 \left(\frac{\left(\chi_{\text{eee}}^{xxx}\right)^2}{\left(1 + jk\chi_{\text{ee}}^{xx}\right)\left(2 - jk\chi_{\text{ee}}^{xx}\right)\left(2 + jk\chi_{\text{ee}}^{xx}\right)^3} \right. \\
& \left. + \frac{\left(\chi_{\text{mmm}}^{yyy}\right)^2}{\left(1 + jk\chi_{\text{mm}}^{yy}\right)\left(2 - jk\chi_{\text{mm}}^{yy}\right)\left(2 + jk\chi_{\text{mm}}^{yy}\right)^3} \right),
\end{aligned}
\tag{4.80a}
$$

$$
\begin{aligned}
E_{\text{bw}}^{\omega} = E_0 & \frac{2jk\left(\chi_{\text{mm}}^{yy} - \chi_{\text{ee}}^{xx}\right)}{\left(2 + jk\chi_{\text{ee}}^{xx}\right)\left(2 + jk\chi_{\text{mm}}^{yy}\right)} \\
& - 4E_0^3 k^2 \left(\frac{\left(\chi_{\text{eee}}^{xxx}\right)^2}{\left(1 + jk\chi_{\text{ee}}^{xx}\right)\left(2 - jk\chi_{\text{ee}}^{xx}\right)\left(2 + jk\chi_{\text{ee}}^{xx}\right)^3} \right. \\
& \left. - \frac{\left(\chi_{\text{mmm}}^{yyy}\right)^2}{\left(1 + jk\chi_{\text{mm}}^{yy}\right)\left(2 - jk\chi_{\text{mm}}^{yy}\right)\left(2 + jk\chi_{\text{mm}}^{yy}\right)^3} \right),
\end{aligned}
\tag{4.80b}
$$

where $k = \omega/c_0$ is the wavenumber at frequency ω. By inspecting this result, we notice that the first term on the right-hand sides correspond to the conventional linear scattering, as already obtained in Eq. (4.23), while the second terms in the brackets correspond to correction terms that take into account the depletion of the pump.

Finally, we provide the nonlinear scattered fields at frequency 2ω but, for conciseness, we only give the contributions from $n = 1$, which read

$$E_{\text{fw}}^{2\omega} = -2jE_0^2 k \left[\frac{\chi_{\text{eee}}^{xxx}}{\left(1 + jk\chi_{\text{ee}}^{xx}\right)\left(2 + jk\chi_{\text{ee}}^{xx}\right)^2} + \frac{\chi_{\text{mmm}}^{yyy}}{\left(1 + jk\chi_{\text{mm}}^{yy}\right)\left(2 + jk\chi_{\text{mm}}^{yy}\right)^2} \right],$$
(4.81a)

$$E_{\text{bw}}^{2\omega} = -2jE_0^2 k \left[\frac{\chi_{\text{eee}}^{xxx}}{\left(1 + jk\chi_{\text{ee}}^{xx}\right)\left(2 + jk\chi_{\text{ee}}^{xx}\right)^2} - \frac{\chi_{\text{mmm}}^{yyy}}{\left(1 + jk\chi_{\text{mm}}^{yy}\right)\left(2 + jk\chi_{\text{mm}}^{yy}\right)^2} \right].$$
(4.81b)

Since we have ignored the contributions from $n = 3$, these expressions correspond to the undepleted pump approximation. If the contributions from $n \geq 3$ were taken into account, then it would be possible to model the depletion of the pump. Note that since the fields in Eq. (4.81) do not incorporate the effects of the depletion of the pump, these expressions are exactly the same as those that would be obtained with the frequency-domain method via Eq. (4.73) provided that $\chi(\omega) = \chi(2\omega)$ [8].

5

Scattered Field Computation

In Chapter 4, we have been interested in finding the susceptibilities of a meta-surface meeting some specifications, either by synthesis, or using the mapping technique between scattering parameters and susceptibilities that was described in Section 4.1.4.

Now, we are interested in computing the scattering response of a metasurface with known susceptibilities. Analyzing the fields scattered by a metasurface may prove particularly useful not only to verify that a metasurface behaves as expected, i.e. as specified by synthesis, but also to predict how it scatters when illuminated by an incident wave that differs from that initially specified in the synthesis procedure.

For computing the fields scattered by an electromagnetic structure, it is common practice to resort to full-wave simulation software. However, there are currently no available simulation tools that include appropriate boundary conditions to satisfactorily model a metasurface. Indeed, as pointed out in Section 3.2 and as modeled in the synthesis methodology of Chapter 4, a metasurface really behaves as a zero-thickness sheet, with possibly both electric and magnetic field discontinuities described by bianisotropic susceptibilities. Although some full-wave simulation algorithm does implement zero-thickness boundary conditions, such as the impedance boundary conditions, they are generally insufficient since they only model electric discontinuities and do not offer the capability to account for bianisotropic susceptibilities. In order to overcome this limitation, we have developed our own simulation tools, which we describe in this chapter. These tools are based on the synthesis methods discussed in Chapter 4 and they are therefore perfectly suited for simulating metasurfaces.

In this chapter, we discuss four different methods to compute the fields scattered by a metasurface. In Section 5.1, we present a method based on Fourier propagation, which is simple but not very accurate for propagation beyond the paraxial limit. In Sections 5.2 and 5.3, we describe two finite-difference methods, one in the frequency domain and the other in the time domain. These methods are rather

Electromagnetic Metasurfaces: Theory and Applications, First Edition.
Karim Achouri and Christophe Caloz.
© 2021 The Institute of Electrical and Electronics Engineers, Inc.
Published 2021 by John Wiley & Sons, Inc.

simple and particularly well-suited for 1D and 2D simulations but do not scale well to 3D simulations. Therefore, we finally present in Section 5.4 an integral equation method that only requires meshing the metasurface and not the entire computational domain, in contrast to the finite-difference methods, and that is thus better suited for 3D simulations.

5.1 Fourier-Based Propagation Method

This method is based on the Fourier propagation technique [49, 115], which allows to compute the electromagnetic fields at any position in space from the knowledge of the fields on a given surface. Here, we consider that the metasurface is lying in the xy-plane at $z = 0$. The electric field at any distance z from the metasurface, $\mathbf{E}(x, y, z)$, is given by the xy two-dimensional Fourier transform pair [115]

$$\mathbf{E}(x, y, z) = \iint_{-\infty}^{+\infty} \hat{\mathbf{E}}(k_x, k_y, z) e^{-j(k_x x + k_y y)} dk_x dk_y, \qquad (5.1a)$$

$$\hat{\mathbf{E}}(k_x, k_y, z) = \frac{1}{4\pi^2} \iint_{-\infty}^{+\infty} \mathbf{E}(x, y, z) e^{j(k_x x + k_y y)} dx dy, \qquad (5.1b)$$

where $\hat{\mathbf{E}}(k_x, k_y, z)$ is the two-dimensional Fourier transform of $\mathbf{E}(x, y, z)$.

Let us assume that we know the electric field right after $(z = 0^+)$ or right before $(z = 0^-)$ the metasurface, i.e. $\mathbf{E}(x, y, 0^\pm)$. We may then compute its Fourier transform, $\hat{\mathbf{E}}(k_x, k_y, 0^\pm)$, using (5.1b) with $z = 0^\pm$. The evolution of the field in the z-direction is then given by [115]

$$\hat{\mathbf{E}}(k_x, k_y, z) = \hat{\mathbf{E}}(k_x, k_y, 0^\pm) e^{\mp j k_z}, \qquad (5.2)$$

with $k_z = \sqrt{k^2 - k_x^2 - k_y^2}$, where the sign of the square root must be chosen so that $\tilde{\mathbf{E}}(k_x, k_y, z) \to 0$ as $z \to \pm\infty$. The electric field at any position z from the metasurface is then obtained by substituting (5.2) into (5.1a), which yields

$$\mathbf{E}(x, y, z) = \iint_{-\infty}^{+\infty} \hat{\mathbf{E}}(k_x, k_y, 0^\pm) e^{-j(k_x x + k_y y \pm k_z z)} dk_x dk_y. \qquad (5.3)$$

However, we still need to know how to find $\mathbf{E}(x, y, 0^\pm)$, from the knowledge of the incident wave and of the metasurface susceptibilities. For this purpose, we may use the scattering matrix approach presented in Section 4.1.4, where the metasurface susceptibilities and the corresponding scattering parameters were related to each other via (4.16). Even though (4.16) was originally derived for the case of a uniform metasurface illuminated by a normally incident plane wave, it applies as a good approximation to the case of paraxial illumination and to a metasurface with slow spatial variations.

For an incident plane wave propagating in the positive z-direction, the fields reflected and transmitted by the metasurface may be, respectively, approximated, using (4.16), by

$$\mathbf{E}_r(x, y, 0^-) \approx \overline{\overline{S}}_{11} \cdot \mathbf{E}_i(x, y, 0), \tag{5.4a}$$

$$\mathbf{E}_t(x, y, 0^+) \approx \overline{\overline{S}}_{21} \cdot \mathbf{E}_i(x, y, 0), \tag{5.4b}$$

while, for an incident wave propagating in the negative z-direction, we have

$$\mathbf{E}_r(x, y, 0^+) \approx \overline{\overline{S}}_{22} \cdot \mathbf{E}_i(x, y, 0), \tag{5.5a}$$

$$\mathbf{E}_t(x, y, 0^-) \approx \overline{\overline{S}}_{12} \cdot \mathbf{E}_i(x, y, 0). \tag{5.5b}$$

The procedure to apply this scattered-field analysis technique thus consists in first computing the reflected and transmitted fields for a given incident field using either (5.4) or (5.5), depending on the illumination direction. These fields are then Fourier transformed using (5.1b) with $z = 0$. Next, they are propagated away from the metasurface along the z-direction using (5.2), and finally inverse Fourier transformed with (5.3) to obtain the final fields.

We now illustrate this Fourier propagation method by an example. Consider a metasurface illuminated by a forward propagating plane wave. The metasurface is synthesized to be reflectionless and to focalize the incident wave into a single point. For simplicity, the problem is restricted to two dimensions, x and z, as shown in Figure 5.1, and all the waves are assumed to be y-polarized.

Before analyzing this scattering problem, we will design the metasurface using the synthesis method described in Section 4.1.2. Since we are considering only y-polarized waves, the metasurface susceptibilities may be obtained with (4.7b) and (4.7c). The incident electric and magnetic fields are, respectively, given by

$$\mathbf{E}_i = \hat{\mathbf{y}} e^{-jkz} \quad \text{and} \quad \mathbf{H}_i = -\hat{\mathbf{x}} \frac{e^{-jkz}}{\eta_0}, \tag{5.6}$$

Figure 5.1 2D metasurface focalizing a normally incident plane wave at a point P of space.

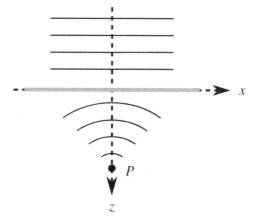

and the reflected fields are $\mathbf{E}_r = \mathbf{H}_r = 0$ since the metasurface is specified to be reflectionless. The transmitted fields are defined as corresponding to that of an electric current line[1] centered at the position of the point P in Figure 5.1. The fields associated with a line source may be obtained from the two-dimensional Green function [31, 67]. Note that, in the literature, a line source is generally defined as being oriented along the z-axis in a cylindrical coordinate system. However, in our case, the line source is oriented along the y-axis, as seen in Figure 5.1. Therefore, we need to perform the cyclic permutation $x \leftrightarrow y', y \leftrightarrow z'$, and $z \leftrightarrow x'$, where the non-primed coordinates correspond to those in Figure 5.1 and the primed ones to those in [31, 67]. In cylindrical coordinates, the two-dimensional Green function is given by [31, 67]

$$g(\rho') = \frac{-jA}{4} H_0^{(2)}(k\rho'), \tag{5.7}$$

where A is the amplitude of the source, $H_0^{(2)}(\cdot)$ is the Hankel function of the second kind of order 0, and ρ' is the radial component of the primed system of coordinates. In Figure 5.1, the coordinates of the point P are $x = 0$ and $z = z_P$, which correspond to $y' = 0$ and $x' = x_p'$. The radial and azimuthal coordinates of the primed system are $\rho' = \sqrt{(x' - x_p')^2 + y'^2}$ and $\phi' = \arctan[y'/(x' - x_p')]$, and correspond to $\rho = \sqrt{(z - z_P)^2 + x^2}$ and $\phi = \arctan[x/(z - z_P)]$. The field radiated by the line source is then found in terms of the Hertz potentials as [31, 67]

$$\mathbf{E} = \nabla(\nabla \cdot \mathbf{\Pi}_e) + k_0^2 \mathbf{\Pi}_e - j\omega\mu_0 \nabla \times \mathbf{\Pi}_m, \tag{5.8a}$$

$$\mathbf{H} = j\omega\epsilon_0 \nabla \times \mathbf{\Pi}_e + \nabla(\nabla \cdot \mathbf{\Pi}_m) + k_0^2 \mathbf{\Pi}_m, \tag{5.8b}$$

where we simply have $\mathbf{\Pi}_e = \hat{\mathbf{z}}'g(\rho')$ and $\mathbf{\Pi}_m = 0$ for a z'-oriented current line. Using (5.7) and (5.8), the transmitted fields tangential to the metasurface are found as

$$\mathbf{E}_t = -\hat{\mathbf{y}}\frac{A}{4}jk_0 H_0^{(2)}(k_0\rho), \tag{5.9a}$$

$$\mathbf{H}_t = \hat{\mathbf{x}}\frac{A}{4}k_0\epsilon_0\omega\cos(\phi)H_1^{(2)}(k_0\rho), \tag{5.9b}$$

where $H_1^{(2)}(\cdot)$ is the Hankel function of the second kind of order 1. Note that the relations (5.9) correspond to the fields *radiated* by a line source, and therefore propagate away from the point P. However, in our synthesis problem, we are interested in the fields converging toward P. This requires reversing the direction of propagation of the fields, which is accomplished by phase conjugation, i.e. by simply flipping the sign of the electric field (5.9a) in this case. Finally, we compute

---

1 We consider a current line since the problem is two dimensional. In the case of a 3D problem, a dipole source should be considered instead.

the metasurface susceptibilities by inserting the specified incident, reflected and transmitted fields into (4.7b) and (4.7c), which yields

$$\chi_{ee}^{yy} = \frac{8 + 2Ak_0\cos(\phi)H_1^{(2)}(k_0\rho)}{4j - Ak_0^2 H_0^{(2)}(k_0\rho)}, \tag{5.10a}$$

$$\chi_{mm}^{xx} = -\frac{2}{k_0}\left(\frac{4j + Ak_0^2 H_0^{(2)}(k_0\rho)}{4 - Ak_0^2\cos(\phi)H_1^{(2)}(k_0\rho)}\right). \tag{5.10b}$$

Now that the metasurface has been synthesized, we can analyze its scattering behavior. For this purpose, we use the Fourier-based scattered-field analysis technique described above. We first verify that the metasurface does indeed behave as a focusing lens. To do so, we determine the fields scattered by the metasurface at $0^{\pm}$ using (5.4). The scattering matrices $\overline{\overline{S}}_{11}$ and $\overline{\overline{S}}_{21}$ are obtained by substituting (5.10) into (4.16), which yields

$$S_{11}^{yy} = \frac{4 + jAk_0^2 H_0^{(2)}(k_0\rho)}{4(k_0 + 1) + Ak_0^2\left[\cos(\phi)H_1^{(2)}(k_0\rho) + jH_0^{(2)}(k_0 r)\right]}$$

$$+ \frac{Ak_0^2 H_0^{(2)}(k_0\rho) + 4j}{Ak_0^2\left[H_0^{(2)}(k_0\rho) - j\cos(\phi)H_1^{(2)}(k_0\rho)\right] + 8j} - 1, \tag{5.11a}$$

$$S_{21}^{yy} = \frac{4 - jAk_0^2 H_0^{(2)}(k_0\rho)}{Ak_0^2\left[\cos(\phi)H_1^{(2)}(k_0\rho) + iH_0^{(2)}(k_0\rho)\right] - 8}$$

$$+ \frac{4 + jAk_0^2 H_0^{(2)}(k_0 r)}{4(k_0 + 1) + Ak_0^2\left[\cos(\phi)H_1^{(2)}(k_0\rho) + iH_0^{(2)}(k_0\rho)\right]}, \tag{5.11b}$$

with all other scattering parameters being zero. For easier visualization, we define the incident field in (5.4) to be a forward-propagating normally incident Gaussian beam, whose electric field is given by $\mathbf{E}_i = \hat{x}\exp[-jkz - x^2]$, instead of a plane wave.

The result is plotted in Figure 5.2a. As can be seen, the metasurface does indeed focalize the incident wave into the specified point, although the focal spot substantially deviates from the fields of a line source. This is because the metasurface has a finite extent along x and is illuminated by a finite size beam, which is different from the infinite extent assumption of (5.9), and because the metasurface is too far away from the focal spot to be able to reconstruct the evanescent field associated with (5.9).

Finally, using the same susceptibilities, we investigate how the metasurface behaves under a different illumination condition. Specifically, we want to know how it scatters when the incident wave propagates at a

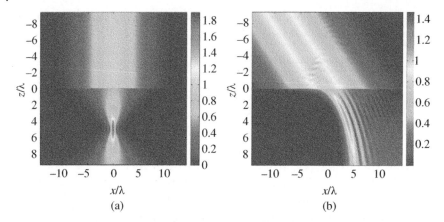

Figure 5.2 Approximate fields scattered by the metasurface when illuminated by (a) a normally incident Gaussian beam and (b) a Gaussian beam incident at 45°. Results obtained for $z_p = 5\lambda$ and $A = 10^{-4}$ for $\omega = 2\pi 10^9$ rad/Hz.

45° angle. This question is answered by using again (5.4) but with $\mathbf{E}_i = \hat{x}\exp[-j\sqrt{2}k(x+z)/2 - \sqrt{2}(z-x)^2/2]$. The result is plotted in Figure 5.2b.

5.2 Finite-Difference Frequency-Domain Method

The Finite-Difference Frequency-Domain (FDFD) method is a rigorous numerical technique to solve Maxwell equations in the frequency domain [142, 162]. Similarly to other conventional electromagnetic finite-difference techniques, the FDFD method consists in discretizing the *spatial* derivatives of Maxwell equation using finite-difference approximation on a staggered Yee grid [162].

The FDFD method considered here uses a conventional computational domain composed of a total-field (TF) region and a scattered-field (SF) region separated by a TF–SF boundary, as illustrated in Figure 5.3. The domain is surrounded at every side by perfectly matched layers (PML), which absorb all electromagnetic radiation. Details about the implementation of the PML and the SF/TF regions are provided in [142].

We will see in the rest of this section how the conventional FDFD equations may be modified to account for the presence of a metasurface. For conciseness, we restrict our attention to the case of a TM_z two-dimensional (xz-plane) problem, where the metasurface may or may not be spatially varying along the x-direction and extends to infinity in the y-direction. For such a problem, the FDFD relations are given by [142]

$$-\frac{H_y^{i,j+1} - H_y^{i,j}}{\Delta z} = j\omega\epsilon_0 \epsilon_{xx}^{i,j+\frac{1}{2}} E_x^{i,j+\frac{1}{2}}, \tag{5.12a}$$

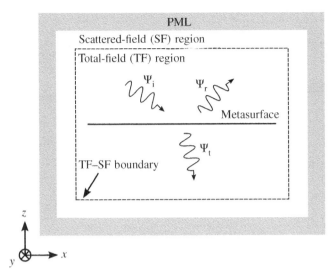

Figure 5.3 FDFD computational domain.

$$\frac{E_x^{i,j+\frac{1}{2}} - E_x^{i,j-\frac{1}{2}}}{\Delta z} - \frac{E_z^{i+\frac{1}{2},j} - E_z^{i-\frac{1}{2},j}}{\Delta x} = j\omega\mu_0\mu_{yy}^{i,j}H_y^{i,j}, \tag{5.12b}$$

$$\frac{H_y^{i+1,j} - H_y^{i,j}}{\Delta x} = j\omega\epsilon_0\epsilon_{zz}^{i+\frac{1}{2},j}E_z^{i+\frac{1}{2},j}, \tag{5.12c}$$

where i and j correspond to the indices of the x and y coordinates of the grid nodes with Δx and Δz being the corresponding lattice periods, as shown in Figure 5.4.

In order to numerically implement relations (5.12), it is practically convenient to express them in the following matrix equations system [142]:

$$-\mathbf{D}_h^z \mathbf{H}_y = \epsilon_{xx}\mathbf{E}_x, \tag{5.13a}$$

$$\mathbf{D}_e^z \mathbf{E}_x - \mathbf{D}_e^x \mathbf{E}_z = \mu_{yy}\mathbf{H}_y, \tag{5.13b}$$

$$\mathbf{D}_h^x \mathbf{H}_y = \epsilon_{zz}\mathbf{E}_z, \tag{5.13c}$$

where $\mathbf{D}_h^z \mathbf{H}_y$ and $\epsilon_{xx}\mathbf{E}_x$ are, respectively,

$$\mathbf{D}_h^z \mathbf{H}_y = \frac{1}{\Delta z}\begin{bmatrix} -1 & 0 & \cdots & 1 & 0 & \cdots & 0 \\ 0 & -1 & 0 & \ddots & 1 & \ddots & 0 \\ 0 & 0 & \ddots & 0 & \ddots & \ddots & 0 \\ \vdots & \ddots & \ddots & -1 & \ddots & 0 & 1 \\ 0 & \ddots & \ddots & \ddots & \ddots & \vdots & \vdots \\ 0 & 0 & \cdots & \cdots & 0 & -1 & 0 \\ 0 & 0 & 0 & \cdots & \cdots & 0 & -1 \end{bmatrix} \cdot \begin{bmatrix} H_y^1 \\ H_y^2 \\ H_y^3 \\ \vdots \\ H_y^{N-2} \\ H_y^{N-1} \\ H_y^N \end{bmatrix}, \tag{5.14}$$

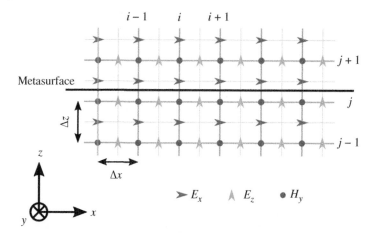

Figure 5.4 FDFD 2D computational domain with a metasurface lying in-between two rows of nodes.

and

$$
\epsilon_{xx}\mathbf{E}_x =
\begin{bmatrix}
\alpha & 0 & 0 & 0 & 0 & \cdots & 0 \\
0 & \alpha & 0 & 0 & 0 & \ddots & 0 \\
0 & 0 & \alpha & 0 & 0 & \ddots & 0 \\
\vdots & \vdots & \ddots & \ddots & \ddots & \ddots & \vdots \\
0 & 0 & 0 & \ddots & \ddots & \ddots & 0 \\
0 & 0 & 0 & \ddots & \ddots & \ddots & 0 \\
0 & 0 & 0 & \cdots & 0 & 0 & \alpha
\end{bmatrix}
\begin{bmatrix}
E_x^1 \\
E_x^2 \\
E_x^3 \\
\vdots \\
E_x^{N-2} \\
E_x^{N-1} \\
E_x^N
\end{bmatrix},
\tag{5.15}
$$

with $\alpha = j\omega\epsilon_0\epsilon_{xx}$ from Eq. (5.12a). In these expressions, the field vectors $\mathbf{E}_x$, $\mathbf{E}_z$ and $\mathbf{H}_y$ contain $N = n_x n_z$ elements, with n_x and n_z being the number of nodes in the x and z directions. The other terms in (5.13) are easily obtained by similarity with (5.14) and (5.15).

In order to simulate a metasurface with this FDFD scheme, one may think that it would be sufficient to assign the metasurface susceptibility values to the FDFD grid cells. However, doing this would cause the two following issues: (i) A grid cell has the lateral dimensions Δx and Δz, which would result in a metasurface with nonzero thickness. This directly conflicts with the assumed zero thickness of the metasurface in the generalized sheet transition conditions (GSTCs). Applying these susceptibilities to a nonzero element would necessarily results in undesired and unpredictable scattering. (ii) A grid cell interconnects either two electric field nodes or two magnetic field nodes, as can be seen in Figure 5.4, which implies that only magnetic or electric field discontinuities may be induced across the metasurface, respectively [166]. This is particularly restricting since a metasurface generally induces both types of discontinuities.

A solution to simulate a metasurface within this scheme is to slightly modify Eqs. (5.13) in accordance with the GSTCs [166, 167]. In order to satisfy the condition of zero thickness, it is necessary to incorporate the metasurface as a virtual entity within the grid by placing it in between electric and magnetic field nodes, as depicted in Figure 5.4. The presence of the metasurface is then taken into account by interconnecting the nodes before and after the metasurface via the GSTCs. For demonstration, consider the case of a bianisotropic metasurface that does not induce rotation of polarization. In the case of a TM$_z$ illumination, the corresponding GSTC equations are

$$-\Delta H_y = j\omega\epsilon_0 \chi_{ee}^{xx} E_{x,av} + jk_0 \chi_{em}^{xy} H_{y,av}, \tag{5.16a}$$

$$-\Delta E_x = j\omega\mu_0 \chi_{mm}^{yy} H_{y,av} + jk_0 \chi_{me}^{yx} E_{x,av}, \tag{5.16b}$$

where the susceptibilities may be varying spatially along the metasurface. Expanding the expressions of the average and difference of the fields in (5.16), we get

$$-H_y^{tr} + H_y^{inc} + H_y^{ref} = j\omega\epsilon_0 \chi_{ee}^{xx} \left(\frac{E_x^{inc} + E_x^{tr} + E_x^{ref}}{2} \right)$$
$$+ jk_0 \chi_{em}^{xy} \left(\frac{H_y^{inc} + H_y^{tr} + H_y^{ref}}{2} \right), \tag{5.17a}$$

$$-E_x^{tr} + E_x^{inc} + E_x^{ref} = j\omega\mu_0 \chi_{mm}^{yy} \left(\frac{H_y^{inc} + H_y^{tr} + H_y^{ref}}{2} \right)$$
$$+ jk_0 \chi_{me}^{yx} \left(\frac{E_x^{inc} + E_x^{tr} + E_x^{ref}}{2} \right). \tag{5.17b}$$

By grouping the incident and reflected fields, we may now reformulate (5.17) using the FDFD notation adopted above, which leads to the following discretized equations:

$$\alpha_m^- H_y^{i,j+1} + \alpha_m^+ H_y^{i,j} = \frac{j\omega\epsilon_0 \chi_{ee}^{xx}}{2} \left(E_x^{i,j+\frac{1}{2}} + E_x^{i,j-\frac{1}{2}} \right), \tag{5.18a}$$

$$\alpha_e^- E_x^{i,j+\frac{1}{2}} + \alpha_e^+ E_x^{i,j-\frac{1}{2}} = \frac{j\omega\epsilon_0 \chi_{mm}^{yy}}{2} \left(H_y^{i,j} + H_y^{i,j+1} \right), \tag{5.18b}$$

where

$$\alpha_m^\pm = \pm 1 - \frac{jk_0 \chi_{em}^{xy}}{2}, \tag{5.19a}$$

$$\alpha_e^\pm = \pm 1 - \frac{jk_0 \chi_{me}^{yx}}{2}. \tag{5.19b}$$

We emphasize here that Eqs. (5.18) are *approximations*. Indeed, in the GSTCs, the susceptibilities are related to the fields just before ($z = 0^-$) and after ($z = 0^+$)

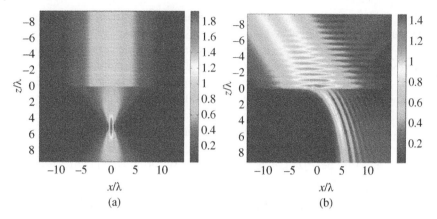

Figure 5.5 FDFD simulations of the fields scattered by a metasurface illuminated by (a) a normally incident Gaussian beam and (b) a Gaussian beam incident at 45°.

the metasurface, while in Eqs. (5.18), the metasurface is in between the nodes meaning that the distance between those before and after the metasurface corresponds to one grid cell (Δz). This introduces an error in the evaluation of the fields scattered by the metasurface, which decreases as the mesh resolution increases. Accordingly, the mesh resolution should be deeply subwavelength, typically in the order of $\lambda/10$, for reasonable accuracy.

We now apply this FDFD scheme to simulate the metasurface problem illustrated in Figure 5.1. The corresponding spatially varying metasurface susceptibilities are substituted in (5.18) and the scattered field is simulated on a $30\lambda \times 18\lambda$ grid with a resolution of $\lambda/30$. As in Figure 5.2, the metasurface is excited by a Gaussian beam at both normal and oblique (45°) incidence. The corresponding simulation results are shown in Figure 5.5a and b, respectively.

A direct comparison between Figures 5.2 and 5.5 reveals that both the Fourier-based and the FDFD methods yield similar results, at least for normal incidence. In the case of oblique incidence, the Fourier-based method leads to undesired effects, as evidenced by the difference in the back scattering and the overall amplitude of the fields. These undesired effects are due to the fact that the Fourier-based method is based on normal scattering parameters, as explained in Section 5.1, while the FDFD method may be applied for any type of illumination conditions.

5.3 Finite-Difference Time-Domain Method

The Finite-Difference Time-Domain (FDTD) method numerically solves Maxwell equations in the time domain. Its principle is similar to that of the FDFD method discussed in the previous section as far as the spatial derivatives are concerned. We do not repeat here the derivation details of the FDTD update equations, as they are largely available in the literature [70, 162]. Instead, we directly consider their most general form, which read

$$
\mathcal{E}_x^{n+\frac{1}{2}}\left(i, j+\frac{1}{2}, k+\frac{1}{2}\right) = \mathcal{E}_x^{n-\frac{1}{2}}\left(i, j+\frac{1}{2}, k+\frac{1}{2}\right)
$$

$$
+ \frac{\Delta t}{\epsilon}\left[\frac{\mathcal{H}_z^n\left(i, j+1, k+\frac{1}{2}\right) - \mathcal{H}_z^n\left(i, j, k+\frac{1}{2}\right)}{\Delta y} \right.
$$

$$
\left. - \frac{\mathcal{H}_y^n\left(i, j+\frac{1}{2}, k+1\right) - \mathcal{H}_y^n\left(i, j+\frac{1}{2}, k\right)}{\Delta z} \right],
$$ (5.20a)

$$
\mathcal{E}_y^{n+\frac{1}{2}}\left(i-\frac{1}{2}, j+1, k+\frac{1}{2}\right) = \mathcal{E}_y^{n-\frac{1}{2}}\left(i-\frac{1}{2}, j+1, k+\frac{1}{2}\right)
$$

$$
+ \frac{\Delta t}{\epsilon}\left[\frac{\mathcal{H}_x^n\left(i-\frac{1}{2}, j+1, k+1\right) - \mathcal{H}_x^n\left(i-\frac{1}{2}, j+1, k\right)}{\Delta z} \right.
$$

$$
\left. - \frac{\mathcal{H}_z^n\left(i, j+1, k+\frac{1}{2}\right) - \mathcal{H}_z^n\left(i-1, j+1, k+\frac{1}{2}\right)}{\Delta x} \right],
$$ (5.20b)

$$
\mathcal{E}_z^{n+\frac{1}{2}}\left(i-\frac{1}{2}, j+\frac{1}{2}, k+1\right) = \mathcal{E}_z^{n-\frac{1}{2}}\left(i-\frac{1}{2}, j+\frac{1}{2}, k+1\right)
$$

$$
+ \frac{\Delta t}{\epsilon}\left[\frac{\mathcal{H}_y^n\left(i, j+\frac{1}{2}, k+1\right) - \mathcal{H}_y^n\left(i-1, j+\frac{1}{2}, k+1\right)}{\Delta x} \right.
$$

$$
\left. - \frac{\mathcal{H}_x^n\left(i-\frac{1}{2}, j+1, k+1\right) - \mathcal{H}_x^n\left(i-\frac{1}{2}, j, k+1\right)}{\Delta y} \right],
$$ (5.20c)

$$
\mathcal{H}_x^{n+1}\left(i-\frac{1}{2}, j+1, k+1\right) = \mathcal{H}_x^n\left(i-\frac{1}{2}, j+1, k+1\right)
$$

$$
+ \frac{\Delta t}{\mu}\left[\frac{\mathcal{E}_y^{n+\frac{1}{2}}\left(i-\frac{1}{2}, j+1, k+\frac{3}{2}\right) - \mathcal{E}_y^{n+\frac{1}{2}}\left(i-\frac{1}{2}, j+1, k+\frac{1}{2}\right)}{\Delta z} \right.
$$

$$
\left. - \frac{\mathcal{E}_z^{n+\frac{1}{2}}\left(i-\frac{1}{2}, j+\frac{3}{2}, k+1\right) - \mathcal{E}_z^{n+\frac{1}{2}}\left(i-\frac{1}{2}, j+\frac{1}{2}, k+1\right)}{\Delta y} \right],
$$ (5.20d)

$$\mathcal{H}_y^{n+1}\left(i, j+\frac{1}{2}, k+1\right) = \mathcal{H}_y^n\left(i, j+\frac{1}{2}, k+1\right)$$

$$+ \frac{\Delta t}{\mu}\left[\frac{\mathcal{E}_z^{n+\frac{1}{2}}\left(i+\frac{1}{2}, j+\frac{1}{2}, k+1\right) - \mathcal{E}_z^{n+\frac{1}{2}}\left(i-\frac{1}{2}, j+\frac{1}{2}, k+1\right)}{\Delta x} \right.$$

$$\left. - \frac{\mathcal{E}_x^{n+\frac{1}{2}}\left(i, j+\frac{1}{2}, k+\frac{3}{2}\right) - \mathcal{E}_x^{n+\frac{1}{2}}\left(i, j+\frac{1}{2}, k+\frac{1}{2}\right)}{\Delta z} \right], \tag{5.20e}$$

$$\mathcal{H}_z^{n+1}\left(i, j+1, k+\frac{1}{2}\right) = \mathcal{H}_z^n\left(i, j+1, k+\frac{1}{2}\right)$$

$$+ \frac{\Delta t}{\mu}\left[\frac{\mathcal{E}_x^{n+\frac{1}{2}}\left(i, j+\frac{3}{2}, k+\frac{1}{2}\right) - \mathcal{E}_x^{n+\frac{1}{2}}\left(i, j+\frac{1}{2}, k+\frac{1}{2}\right)}{\Delta y} \right.$$

$$\left. - \frac{\mathcal{E}_y^{n+\frac{1}{2}}\left(i+\frac{1}{2}, j+1, k+\frac{1}{2}\right) - \mathcal{E}_y^{n+\frac{1}{2}}\left(i-\frac{1}{2}, j+1, k+\frac{1}{2}\right)}{\Delta x} \right], \tag{5.20f}$$

where $(x, y, z) = (i\Delta x, j\Delta y, k\Delta z)$ and $t = n\Delta t$. These equations must be applied according to the leap-frog formulation corresponding to the staggered Yee grid structure depicted in Figure 5.6 for a 2D (2D spatial + 1D time) case. In this formulation, different field components are located at different spatial and temporal positions. Indeed, referring to Eqs. (5.20), we see that the electric field components are sampled at the times $t = (n + \frac{1}{2})\Delta t$, while the magnetic field components are sampled at the times $t = (n + 1)\Delta t$. The procedure in using the update equations (5.20) is thus to first compute the electric field components at a given instant of time. Then, the magnetic fields components may be computed using the previously found electric ones, and so on.

To incorporate a metasurface within the computational domain, we follow a procedure very similar to that used for the FDFD method discussed in Section 5.2, with the additional complexity that the metasurface may be time-varying in addition to being dispersive, i.e. the susceptibilities may be a function of both ω and t. In order to simplify our developments, we treat separately the cases of time-varying dispersionless metasurfaces, which we will discuss in the next section, and that of time-varying dispersive metasurfaces, which will be addressed in Section 5.3.2.

5.3.1 Time-Varying Dispersionless Metasurfaces

A dispersionless metasurface exhibits susceptibilities that do not depend on the frequency, i.e. $\chi \neq \chi(\omega)$. In the time domain, this leads to an important

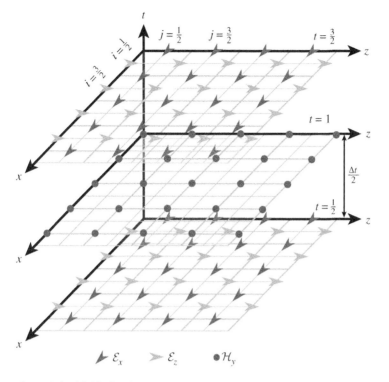

Figure 5.6 2D FDTD grid.

simplification. Indeed, as discussed in Section 2.2, the general expressions of the constitutive relations involve the time-domain convolutions of the susceptibilities with their corresponding electromagnetic fields. However, if the susceptibilities are not functions of the frequency, then these convolutions reduce to simple multiplications, which greatly simplifies the analysis procedure.

For simplicity, we now restrict our attention to 2D FDTD simulations of time-varying dispersionless metasurfaces assuming TM_z illumination. It is straightforward to obtain, from the forthcoming developments, the relations for 1D and 3D simulations as well as for other illumination polarizations. We start by simplifying relations (5.20) for our case study, which yields

$$\mathcal{E}_x^{n+\frac{1}{2}}\left(i, k + \frac{1}{2}\right) = \mathcal{E}_x^{n-\frac{1}{2}}\left(i, k + \frac{1}{2}\right) - \frac{\Delta t}{\epsilon}\left[\frac{\mathcal{H}_y^n(i, k+1) - \mathcal{H}_y^n(i, k)}{\Delta z}\right],$$

$$(5.21a)$$

$$
H_y^{n+1}(i,k) = H_y^n(i,k) + \frac{\Delta t}{\mu}\left[\frac{\mathcal{E}_z^{n+\frac{1}{2}}\left(i+\frac{1}{2},k\right) - \mathcal{E}_z^{n+\frac{1}{2}}\left(i-\frac{1}{2},k\right)}{\Delta x} \right.
$$
$$
\left. - \frac{\mathcal{E}_x^{n+\frac{1}{2}}\left(i,k+\frac{1}{2}\right) - \mathcal{E}_x^{n+\frac{1}{2}}\left(i,k-\frac{1}{2}\right)}{\Delta z} \right], \tag{5.21b}
$$

where we have removed all dependencies to the y coordinate since $\partial_y \to 0$.

As for the FDFD case, we now place the metasurface to simulate at a virtual position in-between rows of electric and magnetic field nodes. For a metasurface lying in the xy-plane, we define the node k_d as the node just before the metasurface with respect to the z-axis, with $k_d + 1$ corresponding to the node right after it. The expressions (5.21) are valid everywhere within the computational domain at the nodes directly surrounding the metasurface, which need to be interconnected via the GSTCs. Consider the electric field $\mathcal{E}_x^{n+\frac{1}{2}}\left(i,k_d+\frac{1}{2}\right)$ given by (5.21a), which depends on $H_y^n(i,k_d+1)$ and $H_y^n(i,k_d)$. Here, $\mathcal{E}_x^{n+\frac{1}{2}}\left(i,k_d+\frac{1}{2}\right)$ and $H_y^n(i,k_d+1)$ are at the same side of the metasurface, and $H_y^n(i,k_d)$ is at the other side. If (5.21a) would be used there without any modification, it would fail to account for the magnetic field discontinuity induced by the metasurface. This discontinuity can be properly accounted for by defining a virtual magnetic field node at $z = 0^+$ such that $H_y^n(i,0^+)$ is used instead of $H_y^n(i,k_d)$. We can then use the GSTCs to connect $H_y^n(i,0^+)$ to the fields on the other side of the metasurface. We can do the same with (5.21b), this time by replacing $\mathcal{E}_x^{n+\frac{1}{2}}\left(i,k+\frac{1}{2}\right)$, which is not on the same side of the metasurface as the other terms in (5.21b), by $\mathcal{E}_x^{n+\frac{1}{2}}(i,0^-)$, where $z = 0^-$ corresponds to the position of the virtual electric field nodes just before the metasurface. We may then rewrite (5.21) for the field nodes right before and after the metasurface as

$$
\mathcal{E}_x^{n+\frac{1}{2}}\left(i,k_d+\frac{1}{2}\right) = \mathcal{E}_x^{n-\frac{1}{2}}\left(i,k_d+\frac{1}{2}\right) - \frac{\Delta t}{\epsilon}\left[\frac{H_y^n(i,k_d+1) - H_y^n(i,0^+)}{\Delta z} \right],
$$
$$
\tag{5.22a}
$$

$$
H_y^{n+1}(i,k_d) = H_y^n(i,k_d) + \frac{\Delta t}{\mu}\left[\frac{\mathcal{E}_z^{n+\frac{1}{2}}\left(i+\frac{1}{2},k_d\right) - \mathcal{E}_z^{n+\frac{1}{2}}\left(i-\frac{1}{2},k_d\right)}{\Delta x} \right.
$$
$$
\left. - \frac{\mathcal{E}_x^{n+\frac{1}{2}}(i,0^-) - \mathcal{E}_x^{n+\frac{1}{2}}\left(i,k_d-\frac{1}{2}\right)}{\Delta z} \right]. \tag{5.22b}
$$

To avoid lengthy equations, we next restrict our attention to homoisotropic metasurfaces with GSTC equations from (3.46), given by

$$\Delta \mathcal{H}_y = -\epsilon_0 \frac{\partial}{\partial t} \left[\tilde{\chi}_{ee}^{xx} \mathcal{E}_{x,av} \right], \tag{5.23a}$$

$$\Delta \mathcal{E}_x = -\mu_0 \frac{\partial}{\partial t} \left[\chi_{mm}^{yy} \mathcal{H}_{y,av} \right], \tag{5.23b}$$

where the time derivatives also apply to the fields and susceptibilities, since both quantities may depend on time. The metasurface is incorporated within the FDTD computational domain upon both space and time discretization. We can now define $\mathcal{H}_y^n \left(i, 0^+ \right)$ and $\mathcal{E}_x^{n+\frac{1}{2}} \left(i, 0^- \right)$ in (5.22) using (5.23), which yields

$$\mathcal{H}_y^n \left(i, 0^+ \right) = \mathcal{H}_y^n \left(i, k_d \right) - \frac{\epsilon_0}{\Delta t} \left[(\tilde{\chi}_{ee}^{xx} \mathcal{E}_{x,av})^{n+\frac{1}{2}} - (\tilde{\chi}_{ee}^{xx} \mathcal{E}_{x,av})^{n-\frac{1}{2}} \right], \tag{5.24a}$$

$$\mathcal{E}_x^{n+\frac{1}{2}} \left(i, 0^- \right) = \mathcal{E}_x^{n+\frac{1}{2}} \left(i, k_d + \frac{1}{2} \right) + \frac{\mu_0}{\Delta t} \left[(\tilde{\chi}_{mm}^{yy} \mathcal{H}_{y,av})^{n+1} - (\tilde{\chi}_{mm}^{yy} \mathcal{H}_{y,av})^n \right]. \tag{5.24b}$$

Substituting (5.24) back into (5.22), replacing $\mathcal{E}_{x,av}^{n-\frac{1}{2}}$ by $\left[\mathcal{E}_x^{n-\frac{1}{2}} \left(i, k_d - \frac{1}{2} \right) + \mathcal{E}_x^{n-\frac{1}{2}} \left(i, k_d + \frac{1}{2} \right) \right]$ and $\mathcal{H}_{y,av}^{n+1}$ by $\left[\mathcal{H}_y^{n+1} \left(i, k_d \right) + \mathcal{H}_y^{n+1} \left(i, k_d + 1 \right) \right]$, and combining identical terms together, results into

$$A_{ee}^{xx,n+\frac{1}{2}} \mathcal{E}_x^{n+\frac{1}{2}} \left(i, k_d + \frac{1}{2} \right)$$
$$= A_{ee}^{xx,n-\frac{1}{2}} \mathcal{E}_x^{n-\frac{1}{2}} \left(i, k_d + \frac{1}{2} \right)$$
$$- \frac{\Delta t}{\epsilon \Delta z} \left[\mathcal{H}_y^n \left(i, k_d + 1 \right) - \mathcal{H}_y^n \left(i, k_d \right) \right]$$
$$+ \frac{1}{2 \Delta z} \left[\tilde{\chi}_{ee}^{xx,n+\frac{1}{2}} \mathcal{E}_x^{n+\frac{1}{2}} \left(i, k_d - \frac{1}{2} \right) - \tilde{\chi}_{ee}^{xx,n-\frac{1}{2}} \mathcal{E}_x^{n-\frac{1}{2}} \left(i, k_d - \frac{1}{2} \right) \right], \tag{5.25a}$$

$$A_{mm}^{yy,n+1} \mathcal{H}_y^{n+1} \left(i, k_d \right)$$
$$= A_{mm}^{yy,n} \mathcal{H}_y^n \left(i, k_d \right) + \frac{\Delta t}{\mu \Delta x} \left[\mathcal{E}_z^{n+\frac{1}{2}} \left(i + \frac{1}{2}, k_d \right) - \mathcal{E}_z^{n+\frac{1}{2}} \left(i - \frac{1}{2}, k_d \right) \right]$$
$$- \frac{\Delta t}{\mu \Delta z} \left[\mathcal{E}_x^{n+\frac{1}{2}} \left(i, k_d + \frac{1}{2} \right) - \mathcal{E}_x^{n+\frac{1}{2}} \left(i, k_d - \frac{1}{2} \right) \right]$$
$$+ \frac{1}{2 \Delta z} \left[\tilde{\chi}_{mm}^{yy,n+1} \mathcal{H}_y^{n+1} \left(i, k_d + 1 \right) - \tilde{\chi}_{mm}^{yy,n} \mathcal{H}_y^n \left(i, k_d + 1 \right) \right], \tag{5.25b}$$

where

$$A_{ee}^{xx,n\pm\frac{1}{2}} = \left(1 + \frac{\tilde{\chi}_{ee}^{xx,n\pm\frac{1}{2}}}{2 \Delta z} \right) \quad \text{and} \quad A_{mm}^{yy,n+1} = \left(1 - \frac{\tilde{\chi}_{mm}^{yy,n+1}}{2 \Delta z} \right). \tag{5.26}$$

These relations are the final FDTD update equations that must be used for the nodes surrounding the metasurface, while (5.21) is to be used everywhere else.

We now illustrate the application of this computational method with an example. Let us consider the case of a time-varying metasurface that exhibits time-dependent absorption. Specifically, the metasurface periodically changes from perfectly absorbing ($T = R = 0$) to half-absorbing ($R = 0$, $T = 0.5$) at a given instant of time. Using (4.22), we find that the frequency-domain susceptibilities for these two cases are, respectively, given by

$$\chi_{ee}^{xx} = \chi_{mm}^{yy} = \frac{2c_0}{j\omega}, \quad \text{and} \quad \chi_{ee}^{xx} = \chi_{mm}^{yy} = \frac{2c_0}{3j\omega}. \tag{5.27}$$

We combine these susceptibilities into the function

$$\tilde{\chi}_{ee}^{xx} = \tilde{\chi}_{mm}^{yy} = \frac{2c_0}{j\omega\alpha(t)} \tag{5.28}$$

where we assume that $\alpha(t)$ has the periodic temporal profile

$$\alpha(t) = \begin{cases} 1 + \dfrac{1}{50\Delta t}t, & \text{for } 0 \le t \le 100\Delta t \\[2mm] 3 - \dfrac{1}{50\Delta t}(t - 100\Delta t), & \text{for } 100\Delta t \le t \le 200\Delta t. \end{cases} \tag{5.29}$$

The susceptibility expression (5.28) presents a very particular property associated with its term $\frac{1}{j\omega}$. Indeed, consider the frequency-domain counterparts of (5.23), which read[2]

$$\Delta H_y = -j\omega\epsilon_0\chi_{ee}^{xx}E_{x,\text{av}}, \tag{5.30a}$$

$$\Delta E_x = -j\omega\mu_0\chi_{mm}^{yy}H_{y,\text{av}}. \tag{5.30b}$$

Upon substituting (5.28) into (5.30), the $j\omega$ terms cancel out, yielding

$$\Delta H_y = -\epsilon_0\frac{2c_0}{\alpha(t)}E_{x,\text{av}}, \tag{5.31a}$$

$$\Delta E_x = -\mu_0\frac{2c_0}{\alpha(t)}H_{y,\text{av}}. \tag{5.31b}$$

Now transforming (5.31) back into the time domain simply gives

$$\Delta \mathcal{H}_y = -\epsilon_0\frac{2c_0}{\alpha(t)}\mathcal{E}_{x,\text{av}}, \tag{5.32a}$$

$$\Delta \mathcal{E}_x = -\mu_0\frac{2c_0}{\alpha(t)}\mathcal{H}_{y,\text{av}}, \tag{5.32b}$$

---

2 Here the tilde indicates frequency-domain fields to avoid confusion with the time-domain ones.

where the time derivatives have vanished when comparing (5.32) to their original form in (5.23). Practically, this implies that the update equations in (5.25) simplify to

$$
A^+ \mathcal{E}_x^{n+\frac{1}{2}}\left(i, k_d + \frac{1}{2}\right) = A^- \mathcal{E}_x^{n-\frac{1}{2}}\left(i, k_d + \frac{1}{2}\right)
$$
$$
- \frac{\Delta t}{\epsilon \Delta z}\left[\mathcal{H}_y^n\left(i, k_d + 1\right) - \mathcal{H}_y^n\left(i, k_d\right)\right]
$$
$$
+ \frac{\Delta t c_0}{2\Delta z \alpha(t)}\left[\mathcal{E}_x^{n+\frac{1}{2}}\left(i, k_d - \frac{1}{2}\right) - \mathcal{E}_x^{n-\frac{1}{2}}\left(i, k_d - \frac{1}{2}\right)\right],
$$

$$(5.33a)$$

$$
A^+ \mathcal{H}_y^{n+1}\left(i, k_d\right) = A^- \mathcal{H}_y^n\left(i, k_d\right)
$$
$$
+ \frac{\Delta t}{\mu \Delta x}\left[\mathcal{E}_z^{n+\frac{1}{2}}\left(i + \frac{1}{2}, k_d\right) - \mathcal{E}_z^{n+\frac{1}{2}}\left(i - \frac{1}{2}, k_d\right)\right]
$$
$$
- \frac{\Delta t}{\mu \Delta z}\left[\mathcal{E}_x^{n+\frac{1}{2}}\left(i, k_d + \frac{1}{2}\right) - \mathcal{E}_x^{n+\frac{1}{2}}\left(i, k_d - \frac{1}{2}\right)\right]
$$
$$
+ \frac{\Delta t c_0}{2\Delta z \alpha(t)}\left[\mathcal{H}_y^{n+1}\left(i, k_d + 1\right) - \mathcal{H}_y^n\left(i, k_d + 1\right)\right],
$$

$$(5.33b)$$

where

$$
A^\pm = \left(1 \pm \frac{\Delta t c_0}{2\Delta z \alpha(t)}\right).
$$

$$(5.34)$$

Let us now illuminate the metasurface with a Gaussian beam with profile $\mathcal{E}_x(x, t) = \sin(2\pi f t)\exp(-x^2/0.02)$, with an excitation frequency of $f = 2$ GHz. The total electric field $\mathcal{E}_x$ obtained after 30 000 time steps with period $\Delta t = 15.349$ ps is plotted in Figure 5.7. The figure clearly shows the expected periodic absorbing behavior, with the transmitted wave amplitude periodically and linearly varying between 0 and 0.5 V/m. Also note that the metasurface is perfectly reflectionless, as synthesized.

5.3.2 Time-Varying Dispersive Metasurfaces

In the previous section, we have derived an FDTD computational scheme that works well but that presents the important limitation that it can only be applied to dispersionless metasurfaces. This is particularly an issue considering that metasurfaces are typically made of resonating scatterers and are hence highly dispersive. To overcome this limitation, we extend here the scheme discussed in Section 5.3.1 to dispersive metasurfaces, i.e. $\chi = \chi(\omega)$, using spectral auxiliary differential equations (ADEs) [162]. This method presents the advantages of being

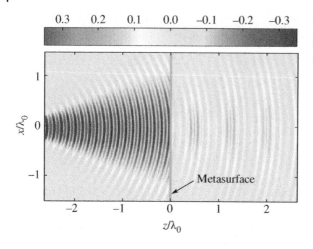

Figure 5.7 FDTD simulated $\mathcal{E}_x$ of a time-varying absorbing metasurface [166].

exact (it does not rely on discretization approximations) and computationally efficient, while still being easily applied to the general case of bianisotropic metasurfaces like the scheme discussed in Section 5.3.1.

For simplicity, we next restrict our attention to 1D FDTD simulations for which the FDTD update equations (5.22), just before and after the metasurface, reduce to

$$\mathcal{E}_x^{n+\frac{1}{2}}\left(k_d + \frac{1}{2}\right) = \mathcal{E}_x^{n-\frac{1}{2}}\left(k_d + \frac{1}{2}\right) - \frac{\Delta t}{\epsilon \Delta z}\left[\mathcal{H}_y^n\left(k_d + 1\right) - \mathcal{H}_y^n\left(0^+\right)\right], \quad (5.35a)$$

$$\mathcal{H}_y^{n+1}\left(k_d\right) = \mathcal{H}_y^n\left(k_d\right) - \frac{\Delta t}{\mu \Delta z}\left[\mathcal{E}_x^{n+\frac{1}{2}}\left(0^-\right) - \mathcal{E}_x^{n+\frac{1}{2}}\left(k_d - \frac{1}{2}\right)\right], \quad (5.35b)$$

where we have again used the concept of virtual nodes. Let us consider a bianisotropic metasurface[3] that only interacts with x-polarized waves, whose frequency-domain GSTC equations are, from (4.2), given by

$$-\Delta H_y = j\omega\epsilon_0 \chi_{ee}^{xx} E_{x,\text{av}} + jk_0 \chi_{em}^{xy} H_{y,\text{av}}, \quad (5.36a)$$

$$-\Delta E_x = j\omega\mu_0 \chi_{mm}^{yy} H_{y,\text{av}} + jk_0 \chi_{me}^{yx} E_{x,\text{av}}, \quad (5.36b)$$

with susceptibilities following the Lorentzian dispersion model, which may be expressed, from (2.23), as

$$\chi(\omega) = \frac{\omega_p^2}{\omega_0^2 - \omega^2 + j2\omega\Gamma}. \quad (5.37)$$

According to the ADE method [162], we now introduce auxiliary bianisotropic functions. Specifically, we found that the following polarization functions are

---

3 We assume here that the metasurface susceptibilities do not vary in time, but this restriction is not essential; it is only a shortness convenience. Note that a time-varying metasurface will be briefly discussed at the end of this section.

appropriate [166]:

$$P_{ee}^{xx} = -j\omega\epsilon_0\chi_{ee}^{xx}E_{x,av}, \tag{5.38a}$$

$$P_{em}^{xy} = -jk_0\chi_{em}^{xy}\tilde{H}_{y,av}, \tag{5.38b}$$

$$M_{mm}^{yy} = -j\omega\mu_0\chi_{mm}^{yy}H_{y,av}, \tag{5.38c}$$

$$M_{me}^{yx} = -jk_0\chi_{me}^{yx}E_{x,av}. \tag{5.38d}$$

Using these expressions, we can rewrite (5.36) as

$$\Delta H_y = P_{ee}^{xx} + P_{em}^{xy}, \tag{5.39a}$$

$$\Delta E_x = M_{mm}^{yy} + M_{me}^{yx}. \tag{5.39b}$$

The terms $\mathcal{H}_y^n(0^+)$ and $\mathcal{E}_x^{n+\frac{1}{2}}(0^-)$ in (5.35) may now be obtained in terms of the auxiliary functions (5.38) by transforming (5.39) into their discrete time-domain counterparts, using inverse Fourier transforms, which yields

$$\mathcal{H}_y^n(0^+) = \mathcal{H}_y^n(k_d) + \frac{P_{ee}^{xx,n+\frac{1}{2}} + P_{ee}^{xx,n-\frac{1}{2}}}{2} + \frac{P_{em}^{xy,n+\frac{1}{2}} + P_{em}^{xy,n-\frac{1}{2}}}{2}, \tag{5.40a}$$

$$\mathcal{E}_x^{n+\frac{1}{2}}(0^-) = \mathcal{E}_x^{n+\frac{1}{2}}\left(k_d + \frac{1}{2}\right) - \frac{M_{mm}^{yy,n+1} + M_{mm}^{yy,n}}{2} - \frac{M_{me}^{yx,n+1} + M_{me}^{yx,n}}{2}. \tag{5.40b}$$

Substituting (5.40) back into (5.35) yields then

$$\mathcal{E}_x^{n+\frac{1}{2}}\left(k_d + \frac{1}{2}\right) = \mathcal{E}_x^{n-\frac{1}{2}}\left(k_d + \frac{1}{2}\right) - \frac{\Delta t}{\epsilon\Delta z}\left[\mathcal{H}_y^n(k_d + 1) - \mathcal{H}_y^n(k_d)\right]$$

$$- \frac{\Delta t}{2\epsilon\Delta z}\left(P_{ee}^{xx,n+\frac{1}{2}} + P_{ee}^{xx,n-\frac{1}{2}} + P_{em}^{xy,n+\frac{1}{2}} + P_{em}^{xy,n-\frac{1}{2}}\right), \tag{5.41a}$$

$$\mathcal{H}_y^{n+1}(k_d) = \mathcal{H}_y^n(k_d) - \frac{\Delta t}{\mu\Delta z}\left[\mathcal{E}_x^{n+\frac{1}{2}}\left(k_d + \frac{1}{2}\right) - \mathcal{E}_x^{n+\frac{1}{2}}\left(k_d - \frac{1}{2}\right)\right]$$

$$+ \frac{\Delta t}{2\mu\Delta z}\left(M_{mm}^{yy,n+1} + M_{mm}^{yy,n} + M_{me}^{yx,n+1} + M_{me}^{yx,n}\right). \tag{5.41b}$$

Note that the first terms on the right-hand sides of these equations correspond to the conventional 1D FDTD update equations, while the second terms correspond to the effect of the metasurface.

The next step is to discretize the auxiliary functions in the time domain. This is accomplished by using (5.37) to transform (5.38a) and (5.38b) into[4]

$$\left(\omega_{0,ee}^2 - \omega^2 + j2\omega\Gamma_{ee}\right)P_{ee}^{xx} = -j\omega\epsilon_0\omega_{p,ee}^2 E_{x,av}, \tag{5.42a}$$

$$\left(\omega_{0,em}^2 - \omega^2 + j2\omega\Gamma_{em}\right)P_{em}^{xy} = -\frac{j\omega}{c_0}\omega_{p,em}^2 H_{y,av}. \tag{5.42b}$$

---

4 Here, we only consider (5.38a) and (5.38b), for simplicity, but similar developments apply to (5.38c) and (5.38d).

We now transform these equations into their time-domain counterparts by replacing $j\omega$ by $\frac{\partial}{\partial t}$ and $-\omega^2$ by $\frac{\partial^2}{\partial t^2}$, which yields

$$\left(\omega_{0,ee}^2 + \frac{\partial^2}{\partial t^2} + 2\Gamma_{ee}\frac{\partial}{\partial t}\right)P_{ee}^{xx} = -\epsilon_0\omega_{p,ee}^2\frac{\partial}{\partial t}\mathcal{E}_{x,av}, \tag{5.43a}$$

$$\left(\omega_{0,em}^2 + \frac{\partial^2}{\partial t^2} + 2\Gamma_{em}\frac{\partial}{\partial t}\right)P_{em}^{xy} = -\frac{\omega_{p,em}^2}{c_0}\frac{\partial}{\partial t}\mathcal{H}_{y,av}. \tag{5.43b}$$

We may discretize these relations using finite differences to obtain

$$\omega_{0,ee}^2 P_{ee}^{xx,n+\frac{1}{2}} + \frac{P_{ee}^{xx,n+\frac{3}{2}} - 2P_{ee}^{xx,n+\frac{1}{2}} + P_{ee}^{xx,n-\frac{1}{2}}}{\Delta t^2}$$

$$+ \Gamma_{ee}\frac{P_{ee}^{xx,n+\frac{3}{2}} - P_{ee}^{xx,n-\frac{1}{2}}}{\Delta t} = -\epsilon_0\omega_{p,ee}^2\frac{\mathcal{E}_{x,av}^{n+\frac{3}{2}} - \mathcal{E}_{x,av}^{n-\frac{1}{2}}}{2\Delta t}, \tag{5.44a}$$

$$\omega_{0,em}^2 P_{em}^{xy,n+\frac{1}{2}} + \frac{P_{em}^{xy,n+\frac{3}{2}} - 2P_{em}^{xy,n+\frac{1}{2}} + P_{em}^{xy,n-\frac{1}{2}}}{\Delta t^2}$$

$$+ \Gamma_{em}\frac{P_{em}^{xy,n+\frac{3}{2}} - P_{em}^{xy,n-\frac{1}{2}}}{\Delta t} = -\frac{\omega_{p,em}^2}{c_0}\frac{\mathcal{H}_{y,av}^{n+1} - \mathcal{H}_{y,av}^{n-1}}{2\Delta t}, \tag{5.44b}$$

where we replace all the n terms by $n-1$ for compatibility with (5.41) and solve the resulting equations for $P_{ee}^{xx,n+\frac{1}{2}}$ and $P_{em}^{xy,n+\frac{1}{2}}$, which yields

$$P_{ee}^{xx,n+\frac{1}{2}} = -\frac{\Delta t^2\omega_{0,ee}^2 - 2}{1 + \Gamma_{ee}\Delta t}P_{ee}^{xx,n-\frac{1}{2}} - \frac{1 - \Gamma_{ee}\Delta t}{1 + \Gamma_{ee}\Delta t}P_{ee}^{xx,n-\frac{3}{2}}$$

$$- \frac{\epsilon_0\Delta t\omega_{p,ee}^2}{2\left(1 + \Gamma_{ee}\Delta t\right)}\left[\mathcal{E}_{x,av}^{n+\frac{1}{2}} - \mathcal{E}_{x,av}^{n-\frac{3}{2}}\right], \tag{5.45b}$$

$$P_{em}^{xy,n+\frac{1}{2}} = -\frac{\Delta t^2\omega_{0,em}^2 - 2}{1 + \Gamma_{em}\Delta t}P_{em}^{xy,n-\frac{1}{2}} - \frac{1 - \Gamma_{em}\Delta t}{1 + \Gamma_{em}\Delta t}P_{em}^{xy,n-\frac{3}{2}}$$

$$- \frac{\Delta t\omega_{p,em}^2}{c_0\left(1 + \Gamma_{em}\Delta t\right)}\left[\mathcal{H}_{y,av}^{n} - \mathcal{H}_{y,av}^{n-2}\right]. \tag{5.45b}$$

We may then obtain the update equation for $\mathcal{E}_x^{n+\frac{1}{2}}\left(k_d + \frac{1}{2}\right)$ by substituting (5.45) into (5.41a), using that $\mathcal{E}_{x,av}^{n+\frac{1}{2}} = \frac{1}{2}\left[\mathcal{E}_x^{n+\frac{1}{2}}(k_d - \frac{1}{2}) + \mathcal{E}_x^{n+\frac{1}{2}}(k_d + \frac{1}{2})\right]$, and solving for $\mathcal{E}_x^{n+\frac{1}{2}}\left(k_d + \frac{1}{2}\right)$. After simplification, we obtain

$$\mathcal{E}_x^{n+\frac{1}{2}}\left(k_d + \frac{1}{2}\right)\left[1 - \frac{\Delta t^2\omega_{p,ee}^2}{8\Delta z\left(1 + \Gamma_{ee}\Delta t\right)}\right] = \mathcal{E}_x^{n-\frac{1}{2}}\left(k_d + \frac{1}{2}\right)$$

$$+ \frac{\Delta t}{\epsilon_0\Delta z}\left[\mathcal{H}_y^n\left(k_d + 1\right) - \mathcal{H}_y^n\left(k_d\right)\right]$$

$$
+ \frac{\Delta t^2 \omega_{p,ee}^2}{4\Delta z \left(1 + \Gamma_{ee}\Delta t\right)} \left[\frac{\mathcal{E}_x^{n+\frac{1}{2}}(k_d - \frac{1}{2})}{2} + \mathcal{E}_{x,av}^{n-\frac{3}{2}} \right]
$$

$$
+ c_1 \mathcal{P}_{ee}^{xx,n-\frac{1}{2}} + c_2 \mathcal{P}_{ee}^{xx,n-\frac{3}{2}} - \frac{\Delta t}{2\epsilon_0 \Delta z} \left(\mathcal{P}_{em}^{xy,n+\frac{1}{2}} + \mathcal{P}_{em}^{xy,n-\frac{1}{2}} \right), \tag{5.46}
$$

where the terms c_1 and c_2 are defined as

$$
c_1 = -\frac{\Delta t}{2\epsilon_0 \Delta z} \left(1 + \frac{\Delta t^2 \omega_{p,ee}^2 + 2}{1 + \Gamma_{ee}\Delta t} \right) \quad \text{and} \quad c_2 = \frac{\Delta t}{2\epsilon_0 \Delta z} \left(\frac{1 - \Gamma_{ee}\Delta t}{1 + \Gamma_{ee}\Delta t} \right). \tag{5.47}
$$

We apply the same methodology to find the expression of the magnetic field update equation, whose magnetic auxiliary functions in (5.41b) may, after discretization of (5.38c) and (5.38d), be expressed as

$$
\mathcal{M}_{mm}^{yy,n+1} = -\frac{\Delta t^2 \omega_{0,mm}^2 - 2}{1 + \Gamma_{mm}\Delta t} \mathcal{M}_{mm}^{yy,n} - \frac{1 - \Gamma_{mm}\Delta t}{1 + \Gamma_{mm}\Delta t} \mathcal{M}_{mm}^{yy,n-1}
$$

$$
- \frac{\mu_0 \Delta t \omega_{p,mm}^2}{2\left(1 + \Gamma_{mm}\Delta t\right)} \left[\mathcal{H}_{y,av}^{n+1} - \mathcal{H}_{y,av}^{n-1} \right], \tag{5.48a}
$$

$$
\mathcal{M}_{me}^{yx,n+1} = -\frac{\Delta t^2 \omega_{0,me}^2 - 2}{1 + \Gamma_{me}\Delta t} \mathcal{M}_{me}^{yx,n} - \frac{1 - \Gamma_{me}\Delta t}{1 + \Gamma_{me}\Delta t} \mathcal{M}_{me}^{yx,n-1}
$$

$$
- \frac{\Delta t \omega_{p,me}^2}{c_0 \left(1 + \Gamma_{me}\Delta t\right)} \left[\mathcal{E}_{x,av}^{n+\frac{1}{2}} - \mathcal{E}_{x,av}^{n-\frac{3}{2}} \right], \tag{5.48b}
$$

and obtain the update equation for $\mathcal{H}_y^{n+1}\left(k_d\right)$ by substituting (5.48) into (5.41b), using the fact that $\mathcal{H}_{y,av}^{n+1} = \frac{1}{2}\left[\mathcal{H}_y^{n+1}(k_d) + \mathcal{H}_y^{n+1}(k_d + 1)\right]$, and solving for $\mathcal{H}_y^{n+1}\left(k_d\right)$. After simplification, we obtain

$$
\mathcal{H}_y^{n+1}\left(k_d\right) \left[1 - \frac{\Delta t^2 \omega_{p,mm}^2}{8\Delta z \left(1 + \Gamma_{mm}\Delta t\right)} \right]
$$

$$
= \mathcal{H}_y^n\left(k_d\right) + \frac{\Delta t}{\mu_0 \Delta z} \left[\mathcal{E}_x^{n+\frac{1}{2}}\left(k_d + \frac{1}{2}\right) - \mathcal{E}_x^{n+\frac{1}{2}}\left(k_d - \frac{1}{2}\right) \right]
$$

$$
+ c_3 \mathcal{M}_{mm}^{yy,n} - \frac{\Delta t^2 \omega_{p,mm}^2}{4\Delta z \left(1 + \Gamma_{mm}\Delta t\right)} \left[\frac{\mathcal{H}_y^{n+1}\left(k_d + 1\right)}{2} - \mathcal{H}_{y,av}^{n-1} \right]
$$

$$
+ c_4 \mathcal{M}_{mm}^{yy,n-1} + \frac{\Delta t}{2\mu_0 \Delta z} \left(\mathcal{M}_{me}^{yx,n+1} + \mathcal{M}_{me}^{yx,n} \right), \tag{5.49}
$$

where the terms c_3 and c_4 are defined as

$$
c_3 = -\frac{\Delta t}{2\epsilon_0 \Delta z} \left(1 + \frac{\Delta t^2 \omega_{p,mm}^2 + 2}{1 + \Gamma_{mm}\Delta t} \right) \quad \text{and} \quad c_4 = \frac{\Delta t}{2\epsilon_0 \Delta z} \left(\frac{1 - \Gamma_{mm}\Delta t}{1 + \Gamma_{mm}\Delta t} \right).
$$

$$\tag{5.50}$$

The final update equations, for the nodes surrounding the metasurface, are thus given in (5.46) and (5.49) for the electric and magnetic fields, respectively, which concludes our analysis of dispersive metasurfaces.

Let us illustrate the application of this computational scheme with an example. We arbitrarily choose the frequency-domain susceptibilities to exhibit the Lorentzian dispersion relations

$$\chi_{ee}^{xx} = \chi_{mm}^{yy} = \frac{1}{\omega_0^2 - \omega^2 + j2\omega\Gamma} \quad \text{and} \quad \chi_{em}^{xy} = \chi_{me}^{yx} = \frac{2}{\omega_0^2 - \omega^2 + j2\omega\Gamma},$$

(5.51)

where we set $\epsilon_0 = \mu_0 = c_0 = 1$, $f = 1$ Hz, $\omega_0 = 40\pi$, and $\Gamma = 8\omega_0$. These susceptibilities essentially correspond to a reflectionless and lossless metasurface, as may be verified by considering a plane wave incident from the left side of the metasurface (port 1), where its reflection (S_{11}) and transmission (S_{21}, at port 2) coefficients are, from (4.16), respectively, given by

$$S_{11} = \frac{2jk_0 \left(\chi_{mm}^{yy} - \chi_{ee}^{xx} + \chi_{me}^{yx} - \chi_{em}^{xy} \right)}{2jk_0 \left(\chi_{ee}^{xx} + \chi_{mm}^{yy} \right) + \chi_{em}^{xy}\chi_{me}^{yx}k_0^2 + 4 - \chi_{ee}^{xx}\chi_{mm}^{yy}k_0^2}.$$

(5.52a)

$$S_{21} = \frac{\chi_{ee}^{xx}\chi_{mm}^{yy}k_0^2 - \left(2j + \chi_{em}^{xy}k_0\right)\left(2j + \chi_{me}^{yx}k_0\right)}{2jk_0 \left(\chi_{ee}^{xx} + \chi_{mm}^{yy} \right) + \chi_{em}^{xy}\chi_{me}^{yx}k_0^2 + 4 - \chi_{ee}^{xx}\chi_{mm}^{yy}k_0^2}.$$

(5.52b)

Substituting (5.51) into (5.52) yields indeed $S_{11} = 0$ and $S_{21} \approx 1$. Using (5.46) and (5.49), we simulate the time-domain response of this metasurface to the Gaussian pulse $\mathcal{E}_x^{inc} = e^{-\left(\frac{t}{\tau}\right)^2}\sin(\omega t)$, where $\tau = 1$. The result is plotted in Figure 5.8, where the spatial position of the source and the metasurface are indicated by the two dots. The vertical dashed lines indicate the transitions

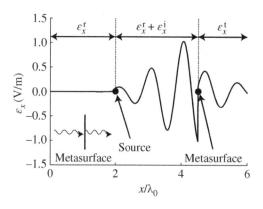

Figure 5.8 FDTD simulated $\mathcal{E}_x$ at $t = 3$ s of a dispersive metasurface with the susceptibilities (5.51).

between the SF regions and TF region.[5] As expected, the reflected wave is zero and almost all power is transmitted through the dispersive metasurface.

Let us finally, for the sake of completeness, consider a *time-varying* dispersive metasurface. In this case, Eq. (5.43) must be modified to account for the time-dependence of the susceptibilities. Assuming that the time variation is due to a time-dependent plasma frequency, i.e. $\omega_{p,ee}(t)$, Eq. (5.43a) becomes

$$\left(\omega_{0,ee}^2 + \frac{\partial^2}{\partial t^2} + 2\Gamma_{ee}\frac{\partial}{\partial t}\right) \mathcal{P}_{ee}^{xx} = -\epsilon_0 \frac{\partial \left(\omega_{p,ee}^2(t) * \mathcal{E}_{x,av}\right)}{\partial t}, \tag{5.53}$$

where a convolution product has been introduced in accordance with the discussion in Section 2.2. To discretize this equation, we may use the technique discussed above for the left-hand side but the right-hand side must be computed using convolutions, which may be achieved as follows:

$$
\begin{aligned}
\frac{\partial \left(\omega_{p,ee}^2(t) * \mathcal{E}_{x,av}\right)}{\partial t} &= \frac{\partial}{\partial t} \int_0^t \omega_{p,ee}^2(t-\tau)\mathcal{E}_{x,av}(\tau)d\tau \\
&= \frac{\int_0^{(n+1)\Delta t} f(n+1,\tau)d\tau - \int_0^{(n-1)\Delta t} f(n-1,\tau)d\tau}{2\Delta t} \\
&= \frac{1}{2\Delta t}\sum_{i=1}^{n} f(n+1,(i+0.5)\Delta t) - f(n-1,(i-0.5)\Delta t),
\end{aligned}
\tag{5.54}
$$

where $f(n,\tau) = \omega_{p,ee}^2(n\Delta t - \tau)\mathcal{E}_{x,av}(\tau)$.

5.4 Spectral-Domain Integral Equation Method

We now present a frequency-domain computational method that may be used to compute the fields scattered by a metasurface by expressing the GSTCs in the spectral domain [30, 68]. The main advantage of this method compared to the finite-difference techniques discussed in Sections 5.2 and 5.3 is that this method requires meshing only the metasurface and not the entire computational domain. Moreover, this method does not require the implementation of PML and is therefore much more numerically efficient, especially for large structures or 3D simulations. Since only the metasurface is meshed, this method also offer the advantage of implementing the GSTCs as exactly zero-thickness transition conditions, which was not the case in the methods of Sections 5.2 and 5.3 since we had to resort to virtual nodes placed at slightly different spatial positions thus introducing errors proportional to the size of the grid cells.

---

5 Not visible here are the two PML regions on both sides of the computational domain.

This method consists in discretizing the GSTCs into a system of spectral-domain integral equations that may then be solved using the method of moments (MoM) [53]. This allows one to compute the scattered fields right before and after the metasurface. To find the fields at any distance from the metasurface, we may then use Fourier-propagation techniques, as in Section 5.1.

Let us start by decomposing in the spectral domain the incident, reflected, and transmitted electric fields. We assume that the metasurface lies in the xy-plane at $z = 0$ and that the incident field propagates in the positive z-direction. In this case, the relationship between the field $\mathcal{E}(\mathbf{r}, t)$ and its spatiotemporal Fourier counterpart[6] $\tilde{\mathbf{E}}(\mathbf{k}_\parallel, \omega)$, where $\mathbf{k}_\parallel = k_x \hat{\mathbf{x}} + k_y \hat{\mathbf{y}}$ and $\mathbf{r} = x\hat{\mathbf{x}} + y\hat{\mathbf{y}}$, may be expressed as

$$\mathcal{E}_i(\mathbf{r}, t) = \int \tilde{\mathbf{E}}_i\left(\mathbf{k}_\parallel, \omega\right) e^{j(\omega t - \mathbf{k}_\parallel \cdot \mathbf{r} - k_z z)} d\omega d\mathbf{k}_\parallel, \tag{5.55a}$$

$$\mathcal{E}_r(\mathbf{r}, t) = \int \tilde{\mathbf{E}}_r\left(\mathbf{k}_\parallel, \omega\right) e^{j(\omega t - \mathbf{k}_\parallel \cdot \mathbf{r} + k_z z)} d\omega d\mathbf{k}_\parallel, \tag{5.55b}$$

$$\mathcal{E}_t(\mathbf{r}, t) = \int \tilde{\mathbf{E}}_t\left(\mathbf{k}_\parallel, \omega\right) e^{j(\omega t - \mathbf{k}_\parallel \cdot \mathbf{r} - k_z z)} d\omega d\mathbf{k}_\parallel, \tag{5.55c}$$

where the integral signs represent

$$\int \cdots d\omega d\mathbf{k}_\parallel = \int_{-\infty}^{+\infty} \int_{-\infty}^{+\infty} \int_{-\infty}^{+\infty} \cdots d\omega dk_x dk_y. \tag{5.56}$$

We may find the corresponding magnetic fields using $\mathbf{H}(\mathbf{r}, \omega) = -j\omega \nabla \times \mathbf{E}(\mathbf{r}, \omega)/\mu_0$ and reducing the differential operator ∇, as

$$\nabla \rightarrow -j\left(\mathbf{k}_\parallel \pm k_z \hat{\mathbf{z}}\right), \tag{5.57a}$$

$$\nabla_\parallel \rightarrow -j\mathbf{k}_\parallel, \tag{5.57b}$$

with $k_z = \pm\sqrt{k_0^2 - k_x^2 - k_y^2}$, where the sign of the square root must satisfy the radiation condition [31, 67], i.e. $Re\{k_z\} \geq 0$ and $Im\{k_z\} \leq 0$ at $z = \pm\infty$. It follows that the magnetic field counterparts of (5.55) read

$$\mathcal{H}_i(\mathbf{r}, t) = \frac{1}{\omega\mu_0} \int \left(\mathbf{k}_\parallel + k_z \hat{\mathbf{z}}\right) \times \tilde{\mathbf{E}}_i\left(\mathbf{k}_\parallel, \omega\right) e^{j(\omega t - \mathbf{k}_\parallel \cdot \mathbf{r} - k_z z)} d\omega d\mathbf{k}_\parallel, \tag{5.58a}$$

$$\mathcal{H}_r(\mathbf{r}, t) = \frac{1}{\omega\mu_0} \int \left(\mathbf{k}_\parallel - k_z \hat{\mathbf{z}}\right) \times \tilde{\mathbf{E}}_r\left(\mathbf{k}_\parallel, \omega\right) e^{j(\omega t - \mathbf{k}_\parallel \cdot \mathbf{r} + k_z z)} d\omega d\mathbf{k}_\parallel, \tag{5.58b}$$

$$\mathcal{H}_t(\mathbf{r}, t) = \frac{1}{\omega\mu_0} \int \left(\mathbf{k}_\parallel + k_z \hat{\mathbf{z}}\right) \times \tilde{\mathbf{E}}_t\left(\mathbf{k}_\parallel, \omega\right) e^{j(\omega t - \mathbf{k}_\parallel \cdot \mathbf{r} - k_z z)} d\omega d\mathbf{k}_\parallel. \tag{5.58c}$$

Let us consider the case of a homoanisotropic metasurface, whose electric and magnetic polarization densities are given by

$$\mathbf{P}(\mathbf{r}, \omega) = \epsilon_0 \overline{\overline{\chi}}_{ee}(\mathbf{r}, \omega) \cdot \mathbf{E}_{av}(\mathbf{r}, \omega), \tag{5.59a}$$

$$\mathbf{M}(\mathbf{r}, \omega) = \overline{\overline{\chi}}_{mm}(\mathbf{r}, \omega) \cdot \mathbf{H}_{av}(\mathbf{r}, \omega), \tag{5.59b}$$

---

6 Here, the tildes indicate (spatial) spectral quantities.

where the susceptibility tensors include both the tangential and normal components, which may be spatially varying and dispersive. These susceptibilities are also expanded in the spectral domain, as

$$\overline{\overline{\chi}}_{ee}(\mathbf{r}, \omega) = \int \overline{\overline{\tilde{\chi}}}_{ee} \left(\mathbf{k}_\parallel, \omega\right) e^{-j\mathbf{k}_\parallel \cdot \mathbf{r}} d\mathbf{k}_\parallel, \tag{5.60a}$$

$$\overline{\overline{\chi}}_{mm}(\mathbf{r}, \omega) = \int \overline{\overline{\tilde{\chi}}}_{mm} \left(\mathbf{k}_\parallel, \omega\right) e^{-j\mathbf{k}_\parallel \cdot \mathbf{r}} d\mathbf{k}_\parallel. \tag{5.60b}$$

Using (5.60), we may now express (5.59) in the spectral domain. This transforms the dot products (5.59) into convolutions, and leads to the tangential components of the polarization densities

$$\tilde{\mathbf{P}}_\parallel \left(\mathbf{k}_\parallel, \omega\right) = \epsilon_0 \int \overline{\overline{\tilde{\chi}}}_{\parallel,ee} \left(\mathbf{k}'_\parallel, \omega\right) \cdot \tilde{\mathbf{E}}_{av} \left(\mathbf{k}_\parallel - \mathbf{k}'_\parallel, \omega\right) d\mathbf{k}'_\parallel, \tag{5.61a}$$

$$\tilde{\mathbf{M}}_\parallel \left(\mathbf{k}_\parallel, \omega\right) = \int \overline{\overline{\tilde{\chi}}}_{\parallel,mm} \left(\mathbf{k}'_\parallel, \omega\right) \cdot \tilde{\mathbf{H}}_{av} \left(\mathbf{k}_\parallel - \mathbf{k}'_\parallel, \omega\right) d\mathbf{k}'_\parallel, \tag{5.61b}$$

while their normal components read

$$\tilde{P}_z \left(\mathbf{k}_\parallel, \omega\right) = \epsilon_0 \int \overline{\overline{\tilde{\chi}}}_{\perp,ee} \left(\mathbf{k}'_\parallel, \omega\right) \cdot \tilde{\mathbf{E}}_{av} \left(\mathbf{k}_\parallel - \mathbf{k}'_\parallel, \omega\right) d\mathbf{k}'_\parallel, \tag{5.62a}$$

$$\tilde{M}_z \left(\mathbf{k}_\parallel, \omega\right) = \int \overline{\overline{\tilde{\chi}}}_{\perp,mm} \left(\mathbf{k}'_\parallel, \omega\right) \cdot \tilde{\mathbf{H}}_{av} \left(\mathbf{k}_\parallel - \mathbf{k}'_\parallel, \omega\right) d\mathbf{k}'_\parallel. \tag{5.62b}$$

In these relations, the susceptibility tensors are given by

$$\overline{\overline{\tilde{\chi}}}_{\parallel,ee} \left(\mathbf{k}_\parallel, \omega\right) = \begin{bmatrix} \tilde{\chi}_{ee}^{xx} \left(\mathbf{k}_\parallel, \omega\right) & \tilde{\chi}_{ee}^{xy} \left(\mathbf{k}_\parallel, \omega\right) & \tilde{\chi}_{ee}^{xz} \left(\mathbf{k}_\parallel, \omega\right) \\ \tilde{\chi}_{ee}^{yx} \left(\mathbf{k}_\parallel, \omega\right) & \tilde{\chi}_{ee}^{yy} \left(\mathbf{k}_\parallel, \omega\right) & \tilde{\chi}_{ee}^{yz} \left(\mathbf{k}_\parallel, \omega\right) \end{bmatrix}, \tag{5.63a}$$

$$\overline{\overline{\tilde{\chi}}}_{\parallel,mm} \left(\mathbf{k}_\parallel, \omega\right) = \begin{bmatrix} \tilde{\chi}_{mm}^{xx} \left(\mathbf{k}_\parallel, \omega\right) & \tilde{\chi}_{mm}^{xy} \left(\mathbf{k}_\parallel, \omega\right) & \tilde{\chi}_{mm}^{xz} \left(\mathbf{k}_\parallel, \omega\right) \\ \tilde{\chi}_{mm}^{yx} \left(\mathbf{k}_\parallel, \omega\right) & \tilde{\chi}_{mm}^{yy} \left(\mathbf{k}_\parallel, \omega\right) & \tilde{\chi}_{mm}^{yz} \left(\mathbf{k}_\parallel, \omega\right) \end{bmatrix}, \tag{5.63b}$$

$$\overline{\overline{\tilde{\chi}}}_{\perp,ee} \left(\mathbf{k}_\parallel, \omega\right) = \begin{bmatrix} \tilde{\chi}_{ee}^{zx} \left(\mathbf{k}_\parallel, \omega\right) & \tilde{\chi}_{ee}^{zy} \left(\mathbf{k}_\parallel, \omega\right) & \tilde{\chi}_{ee}^{zz} \left(\mathbf{k}_\parallel, \omega\right) \end{bmatrix}, \tag{5.63c}$$

$$\overline{\overline{\tilde{\chi}}}_{\perp,mm} \left(\mathbf{k}_\parallel, \omega\right) = \begin{bmatrix} \tilde{\chi}_{mm}^{zx} \left(\mathbf{k}_\parallel, \omega\right) & \tilde{\chi}_{mm}^{zy} \left(\mathbf{k}_\parallel, \omega\right) & \tilde{\chi}_{mm}^{zz} \left(\mathbf{k}_\parallel, \omega\right) \end{bmatrix}. \tag{5.63d}$$

Finally, we also need to express in the spectral domain the $\hat{\mathbf{z}} \times \nabla_\parallel \mathcal{M}_z$ and $\hat{\mathbf{z}} \times \nabla_\parallel P_z$ terms of the general GSTC Eqs. (4.1a) and (4.1b). Using (5.57), these terms may be expressed as

$$\hat{\mathbf{z}} \times \nabla_\parallel P_z(\mathbf{r}, t) = \int \left(\hat{\mathbf{z}} \times \mathbf{k}_\parallel\right) \tilde{P}_z \left(\mathbf{k}_\parallel, \omega\right) e^{j(\omega t - \mathbf{k}_\parallel \cdot \mathbf{r})} d\omega d\mathbf{k}_\parallel, \tag{5.64a}$$

$$\hat{\mathbf{z}} \times \nabla_\parallel \mathcal{M}_z(\mathbf{r}, t) = \int \left(\hat{\mathbf{z}} \times \mathbf{k}_\parallel\right) \tilde{M}_z \left(\mathbf{k}_\parallel, \omega\right) e^{j(\omega t - \mathbf{k}_\parallel \cdot \mathbf{r})} d\omega d\mathbf{k}_\parallel. \tag{5.64b}$$

The final form of the spectral-domain integral equations are obtained by substituting (5.55), (5.58), (5.61), (5.62), and (5.64) into (4.1a) and (4.1b). After some manipulations, Eq. (4.1a) becomes

$$
\frac{1}{\omega\mu_0}\hat{z} \times \left[- \left(\mathbf{k}_\parallel + k_z\hat{z}\right) \times \tilde{\mathbf{E}}_i\left(\mathbf{k}_\parallel, \omega\right) - \left(\mathbf{k}_\parallel - k_z\hat{z}\right) \times \tilde{\mathbf{E}}_r\left(\mathbf{k}_\parallel, \omega\right)\right.
$$

$$
\left. + \left(\mathbf{k}_\parallel + k_z\hat{z}\right) \times \tilde{\mathbf{E}}_t\left(\mathbf{k}_\parallel, \omega\right)\right] = \frac{1}{2}j\omega\epsilon_0 \int \overline{\overline{\chi}}_{\parallel,ee}\left(\mathbf{k}_\parallel - \mathbf{k}'_\parallel, \omega\right)
$$

$$
\cdot \left[\tilde{\mathbf{E}}_i\left(\mathbf{k}'_\parallel, \omega\right) + \tilde{\mathbf{E}}_r\left(\mathbf{k}'_\parallel, \omega\right) + \tilde{\mathbf{E}}_t\left(\mathbf{k}'_\parallel, \omega\right)\right] d\mathbf{k}'_\parallel
$$

$$
- \frac{1}{2\omega\mu_0} \int \hat{z} \times \left(\mathbf{k}'_\parallel - \mathbf{k}_\parallel\right) \overline{\overline{\chi}} \perp, mm \left(\mathbf{k}'_\parallel - \mathbf{k}_\parallel, \omega\right)
$$

$$
\cdot \left\{ \left(\mathbf{k}'_\parallel + k'_z\hat{z}\right) \times \left[\tilde{\mathbf{E}}_i\left(\mathbf{k}'_\parallel, \omega\right) + \tilde{\mathbf{E}}_t\left(\mathbf{k}'_\parallel, \omega\right)\right]\right.
$$

$$
\left. + \left(\mathbf{k}'_\parallel - k'_z\hat{z}\right) \times \tilde{\mathbf{E}}_r\left(\mathbf{k}'_\parallel, \omega\right)\right\} d\mathbf{k}'_\parallel, \tag{5.65}
$$

while (4.1b) becomes

$$
\hat{z} \times \left[-\tilde{\mathbf{E}}_i\left(\mathbf{k}_\parallel, \omega\right) - \tilde{\mathbf{E}}_r\left(\mathbf{k}_\parallel, \omega\right) + \tilde{\mathbf{E}}_t\left(\mathbf{k}_\parallel, \omega\right)\right]
$$

$$
= -\frac{1}{2}\mu_0 j\omega \int \frac{1}{\omega\mu_0}\overline{\overline{\chi}}_{\parallel,mm}\left(\mathbf{k}_\parallel - \mathbf{k}'_\parallel, \omega\right)
$$

$$
\cdot \left[\left(\mathbf{k}'_\parallel + k'_z\hat{z}\right) \times \tilde{\mathbf{E}}_i\left(\mathbf{k}'_\parallel, \omega\right) + \left(\mathbf{k}'_\parallel - k'_z\hat{z}\right)\right.
$$

$$
\left. \times \tilde{\mathbf{E}}_r\left(\mathbf{k}'_\parallel, \omega\right) + \left(\mathbf{k}'_\parallel + k'_z\hat{z}\right) \times \tilde{\mathbf{E}}_t\left(\mathbf{k}'_\parallel, \omega\right)\right] d\mathbf{k}'_\parallel
$$

$$
- \frac{1}{2\epsilon_0}\epsilon_0 \int \hat{z} \times \left(\mathbf{k}'_\parallel - \mathbf{k}_\parallel\right) \overline{\overline{\chi}}_{\perp,ee}\left(\mathbf{k}'_\parallel - \mathbf{k}_\parallel, \omega\right)
$$

$$
\cdot \left[\tilde{\mathbf{E}}_i\left(\mathbf{k}'_\parallel, \omega\right) + \tilde{\mathbf{E}}_r\left(\mathbf{k}'_\parallel, \omega\right) + \tilde{\mathbf{E}}_t\left(\mathbf{k}'_\parallel, \omega\right)\right] d\mathbf{k}'_\parallel. \tag{5.66}
$$

The procedure to solve the system of integral equations formed by the combination of (5.65) and (5.66) consists first in transforming $\mathcal{E}_i(\mathbf{r}, t)$, $\overline{\overline{\chi}}_{ee}(\mathbf{r}, t)$, and $\overline{\overline{\chi}}_{mm}(\mathbf{r}, t)$, which are the known quantities of this problem, into their spectral-domain counterparts $\tilde{\mathbf{E}}_i(\mathbf{k}_\parallel, \omega)$, $\overline{\overline{\chi}}_{ee}(\mathbf{k}_\parallel, \omega)$, and $\overline{\overline{\chi}}_{mm}(\mathbf{k}_\parallel, \omega)$, respectively. Then, these equations are solved to provide the spectral-domain reflected and transmitted electric fields $\tilde{E}_{r,x}(\mathbf{k}_\parallel, \omega)$, $\tilde{E}_{r,y}(\mathbf{k}_\parallel, \omega)$, $\tilde{E}_{t,y}(\mathbf{k}_\parallel, \omega)$, and $\tilde{E}_{t,y}(\mathbf{k}_\parallel, \omega)$. Finally, these fields are transformed back into direct space using (5.55). As previously mentioned, we may use the MoM technique to solve for the four unknown electric field components. For this purpose, we expand them into a set of basis functions $B_n\left(\mathbf{k}_\parallel\right)$, as

$$
\tilde{E}_{r/t,x/y}\left(\mathbf{k}_\parallel, \omega\right) = \sum a_n^{r/t,x/y} B_n\left(\mathbf{k}_\parallel\right), \tag{5.67}
$$

where $a_n^{r/t,x/y}$ represents the unknown coefficients corresponding to the different field components. Next, we transform (5.65) and (5.66) into a system of linear equation by multiplication with the test function $W_n(\mathbf{k}_\parallel)$ and integrating the resulting set of equations over $\mathbf{k}_\parallel$. In what follows, we use the point matching technique, i.e. $W_n(\mathbf{k}_\parallel) = \delta(\mathbf{k}_\parallel - \mathbf{k}_{n\parallel})$ and Galerkin testing, i.e. $B_n(\mathbf{k}_\parallel) = W_n(\mathbf{k}_\parallel)$ [68].

Let us consider an example to illustrate the application of this spectral-domain integral method. We consider a reflectionless polychromatic focusing metasurface, which focalizes four incident plane waves, with wavelengths λ_1, λ_2, λ_3 and λ_4, into four distinct focal points. Assuming that the metasurface is homoisotropic with only tangential susceptibility components, the corresponding GSTCs read

$$\Delta H_y = -j\omega\epsilon_0 \chi_{ee} E_{x,\text{av}}, \tag{5.68}$$

$$\Delta E_x = -j\omega\mu_0 \chi_{mm} H_{y,\text{av}}, \tag{5.69}$$

where the fields on the left-hand side ($z = 0^-$) of the metasurface are specified as x-polarized plane waves and those on its right-hand side ($z = 0^+$) are defined using the Green function (5.7). The four focal points are placed at positions $z = \lambda_1$ and $x_1 = -2\lambda_1, x_2 = -0.66\lambda_1, x_3 = 0.66\lambda_1$, and $x_4 = 2\lambda_1$. The four wavelengths correspond to equidistant frequencies ranging from ω_1 to $2\omega_1$. The susceptibility synthesis is performed separately for each of the four wavelength and the resulting susceptibilities are plotted in Figure 5.9.

We now use the spectral-domain integral equation method to compute the fields scattered by the synthesized metasurface. For this purpose, we substitute the susceptibilities plotted in Figure 5.9 into the system formed by Eqs. (5.65) and (5.66) and solve this system for the scattered fields. For an incident field composed of the sum of the four wavelengths, the resulting scattered field is plotted in Figure 5.10,

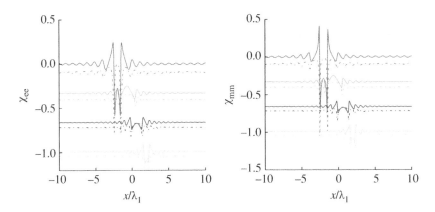

Figure 5.9 Electric and magnetic susceptibilities of the polychromatic focusing metasurface. For better visualization, the three bottom curves have been vertically offset, compared to the top one, by −0.33, −0.66, and −1.0 U, respectively [30].

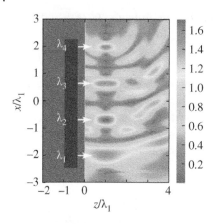

Figure 5.10 Amplitude of the reflected and transmitted fields scattered by the polychromatic metasurface when illuminated by a sum of plane waves at wavelength λ_1, λ_2, λ_3 and λ_4 [30].

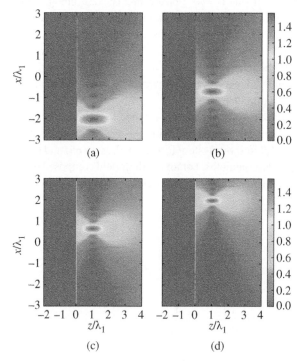

Figure 5.11 Amplitude of the reflected and transmitted fields scattered by the polychromatic metasurface when illuminated by a plane wave at: (a) λ_1, (b) λ_2, (c) λ_3, and (d) λ_4 [30].

where 256 basis functions have been used. As clearly visible, the reflected field is zero, as specified. It may a priori seem that the transmitted field does not correspond to the expected focal spots. However, this is an artifact due to the superposition of the fields at different wavelengths, as may be easily verified upon exciting the metasurface at each wavelength separately, as done in Figure 5.11.

6

Practical Implementation

This chapter presents different strategies to practically implement a metasurface. Specifically, it explains how to bridge the gap between a metasurface described in terms of susceptibility tensors, following the synthesis procedure of Chapter 4, and an actual metasurface made of physical scattering particles.

For this purpose, Section 6.1 starts by describing the general metasurface practical implementation procedure and related concepts, such as the discretization of the susceptibility functions and the simulation of the electromagnetic response of scattering particles. Then, Section 6.2 deals with the design of metallic or dielectric scattering particles to control the phase of the transmitted and reflected waves and explains the related induction of electric and magnetic dipolar responses. This is done for simple scattering particles, such as thin metallic patches, which can achieve full-phase scattering coverage (0 to 2π) but suffer from limited control of the scattered waves amplitude. To overcome this limitation, Section 6.3 presents scattering particles composed of cascaded metallic layers, which offer both amplitude and phase control. Finally, Section 6.4 presents the more advanced topic of the effects of structural symmetry breaking on the angular scattering and polarization responses in a metasurface.

6.1 General Implementation Procedure

Chapter 4 explained the determination of the continuous susceptibility tensorial functions corresponding to a specified metasurface transformation. This represents the first step of the synthesis procedure. The second step consists in discretizing these susceptibility functions into unit cells. The electromagnetic response of each of these unit cells is then physically implemented with scattering particles of proper material and geometry. The third, and last, step of the synthesis procedure thus consists in finding the shapes, dimensions, orientations, and materials of the scattering particles that realize the desired susceptibilities.

Electromagnetic Metasurfaces: Theory and Applications, First Edition.
Karim Achouri and Christophe Caloz.
© 2021 The Institute of Electrical and Electronics Engineers, Inc.
Published 2021 by John Wiley & Sons, Inc.

As discussed in Section 3.1.4, the discretization[1] of the susceptibility functions is to be performed so that the unit cell dimension, in the metasurface plane, is less than $\lambda_0/(n_1 + n_2)$, where n_1 and n_2 are the refractive indices of the media before and after the metasurface. For a uniform metasurface, for which $\chi \neq \chi(x,y)$, the exact dimension of the unit cell is not very important as long as it is smaller than the above limit. For a metasurface with periodic susceptibility functions, then the unit cell size, d, should be such that $P = Nd$, where N is an integer and P is the period of the susceptibility functions either in the x- or y-direction. Here, d should again be less than the above limit and $N \geq 2$ to satisfy the Nyquist–Shannon sampling criterion [149]. Finally, for a spatially varying but nonperiodic metasurface, the unit cell size should be smaller than the above limit and also small enough to satisfy the Nyquist–Shannon sampling criterion.

The strategy for designing the scattering particles essentially consists in full-wave simulating the scattering coefficients of particles with presumably appropriate features. The simulations are first performed over a relatively broad range of frequencies (or wavelengths) to ensure that the resonant frequency of the particles is near the desired frequency of operation. Then, the shape of the particles is tuned across a reasonable range of parameter values, forming the entries of a parametric table. Next, the scattering coefficient for each unit cell of the metasurface is computed from the synthesized susceptibilities using the mapping functions (4.11) and (4.16). Finally, the corresponding scattering particle parameters are collected from the table.

Determining the exact scattering parameters of the metasurface scattering particles requires the use of a 3D full-wave simulation software. A typical procedure consists in separately simulating the different scattering particles within periodic (or Floquet) boundary conditions to model or approximate the effects of periodicity or local quasiperiodicity of the metasurface, taking into account the electromagnetic coupling between adjacent particles. In the case of a nonuniform metasurface, this approach induces some errors. However, these errors are typically negligible when the metasurface spatial variations are small compared to the wavelength, and can be eliminated by further optimization.

Figure 6.1 shows the schematic of a typical simulation setup for a metasurface unit cell. The scattering particle is represented here as an electrically thin metallic patch deposited on a dielectric substrate.[2] Two sets of periodic boundary

---

1 We assume that the metasurface is discretized into a square or rectangular lattice.
2 In the optical regime, the substrate on which the metasurface scattering particles are deposited is typically several hundreds or even thousands of wavelengths thick for sufficient mechanical strength to support the weight of the metasurface so that it does not break. The entire substrate is thus not simulated, as this would require too much computational resources. Instead, the substrate is treated as a half-space medium, as shown in Figure 6.1. This approach introduces some errors since it is unable to account for the light escaping the substrate.

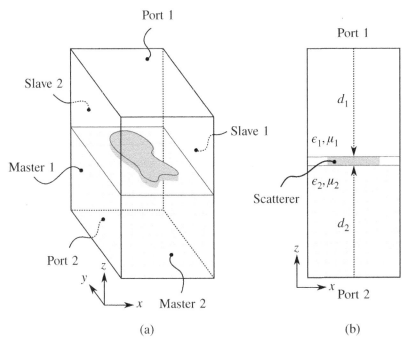

Figure 6.1 Full-wave simulation setup showing a metasurface unit cell with a metallic scatterer deposited on a dielectric substrate with parameters ϵ_2 and μ_2 and background medium with parameters ϵ_1 and μ_1. (a) 3D view of the computational domain. (b) Side view showing the de-embedding distances d_1 and d_2.

conditions usually relate the edges of the periodic cell in each direction. The side denoted Master 1 is related to Slave 1, and the one denoted Master 2 is related Slave 2. Finally, a pair of exciting/receiving ports is placed on the top and bottom sides of the computational domain.[3]

The distance between the ports and the scatterer should be chosen judiciously. If the distance is too short, the near-fields produced by the scatterer interact with the ports, which introduces numerical errors. On the other hand, if the ports are too far away, the computational time is excessive. A good rule of thumb is to place the

---

However, for a transparent substrate like glass, the reflection at the output glass–air interface is only $\approx 3\%$, which is, in most cases, negligible. In the microwave regime, the metasurface substrate can be subwavelengthly thick and mechanically strong enough to support its own weight. In that case, the entire substrate is usually accounted for in the computational domain.
3 Note that in some simulation software, it is also required to place perfectly matched layers (PML) beyond each port to absorb the electromagnetic energy and thus avoid spurious scattering.

ports at about three wavelengths away from the scatterers, where the evanescent near-fields have decayed to a negligible level.

As explained at the beginning of this section, special care should be taken in handling the period of the unit cell. One should not only ensure a sufficient sampling of the susceptibility functions but also avoid the excitation of diffraction orders. If diffraction orders appear in either of the two media in Figure 6.1, the simulation results cannot be trusted as these diffraction orders are usually not properly accounted for by the exciting/receiving ports that spuriously scatter them, leading to substantial errors.

Finally, a de-embedding operation is generally required to ensure that the phase information provided by the scattering parameters exclusively corresponds to the phase shift induced by the scatterer. This de-embedding operation[4] consists in removing the phase that the incident and scattered waves have accumulated in their propagation in the top and bottom media. This accumulated phase may be removed by multiplying the scattering parameters by $\exp[jk_0(\pm d_i n_i \pm d_s n_s)]$, where $d_{i/s}$ and $n_{i/s}$ represent the distance traveled by the incident/scattered waves from the ports to the top and bottom sides of the scatterer, as depicted by the two arrows in Figure 6.1b, and the refractive index of the corresponding medium, respectively. The $\pm$ signs should be chosen according to the direction of wave propagation and the harmonic time-dependence used by the simulation software.

6.2 Basic Strategies for Full-Phase Coverage

In the following, we shall consider unit cells whose lateral dimensions are smaller than the wavelength so as to prevent the excitation of diffraction orders. The scattering response of the metasurface is then essentially due to induced electric and magnetic dipolar modes, as modeled by the GSTCs. It will then be necessary to design scattering particles that exhibit both electric and magnetic dipolar responses and that have sufficient degrees of freedom to realize the required dipolar responses for the practical implementation of metasurfaces.

In general, a metasurface scattering particle must be able to control the amplitude, phase, and polarization of the scattered waves. Controlling the amplitude may be achieved using either active or dissipative elements. In the microwave regime, they may be implemented using chip amplifiers and resistive materials. In the optical regime, gain is generally not implemented, for practical reasons, although gain media exist, while dissipation may be realized using lossy materials such as metals. Controlling the phase of the scattered waves is most important in all metasurface transformations that involve a change in the direction of wave

---

4 This operation is automated in some full-wave simulation software.

propagation. Note that due to periodicity, it is sufficient to induce a phase shift within the $0-2\pi$ range to fully control the phase of the scattered waves. While there exist many possible strategies to accomplish such a phase control, we will next concentrate on the most common ones, which apply to linearly and circularly polarized waves and to both metallic and dielectric scattering particles. Finally, strategies for controlling the polarization will be discussed in Section 6.4.2.

Note that in this section, we will be discussing scattering particles with relatively simple geometries. Such particles are therefore the best candidates for metasurfaces operating in the optical regime. Scattering particles with more complicated shapes have more advanced field control capabilities but they are mostly restricted by fabrication to the microwave regime. They will be discussed in Section 6.3.

6.2.1 Linear Polarization

We shall now present basic strategies to achieve a full-phase coverage in the case of linearly polarized waves for both metallic and dielectric scattering particles.

6.2.1.1 Metallic Scattering Particles

As explained at the beginning of this section, a scattering particle should ideally exhibit both electric and magnetic dipolar responses, and should be able to induce a phase-shift covering a range of 2π, while maintaining a high scattering amplitude for efficient wave manipulation. We shall explain the relationships between these requirements and show how they can be accommodated with metallic scattering particles. For this purpose, we will first investigate how an electric dipolar response may be induced with a metallic scattering particle and demonstrate that a lack of magnetic dipolar response results in a limited scattered wave phase coverage. To overcome this limitation, we will then see how both electric and magnetic dipolar responses may be induced with a metallic scattering particle and show that it indeed leads to full-phase coverage.

To illustrate these two cases, we will consider two different metasurface designs that are conventionally used. The first one is based on transparent scattering particles that may be used both in reflection and in transmission. The second one is based on impenetrable scattering particles that can thus only be used in reflection. Therefore, to fairly compare the scattering response of these two metasurface designs, we shall next concentrate our attention on the reflection properties of both designs. However, note that the concepts that we shall present also apply to the case of transmission metasurfaces.

A well-known technique to obtain an electric dipole is to use a small metallic patch or rod [67]. Controlling the scattering response of such a structure can then be achieved by tuning its length and width. The simplest way to realize such

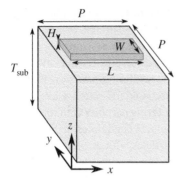

Figure 6.2 Simple metasurface scattering particle consisting of a thin metallic patch on a dielectric substrate.

a scattering particle is to deposit a subwavelength metallic patch on a dielectric substrate, as shown in Figure 6.2.

We next assume a metasurface design in the optical regime at an operation wavelength of 1064 nm. We consider that the metallic patch is made of gold and the substrate of SiO$_2$. Following the procedure in Section 6.1, we first heuristically set initial values for the dimensions of the structure and, to identify its resonant wavelength, compute its reflection and transmission coefficients across the wavelength range from 700 to 1700 nm using a full-wave simulation software. Then, we tune the dimensions of the structure until it resonates near the desired operation wavelength of 1064 nm. The resulting scattered power and de-embedded phase, for an x-polarized illumination,[5] are plotted in Figure 6.3a and b, respectively.

As can be seen, the structure indeed exhibits a strong resonance near the desired operation wavelength, but suffers from the two following limitations: the amplitude of its reflection coefficient quickly decays away from the resonance and the corresponding phase shift range barely reaches π, which is only half of the desired range. The decay in the scattering amplitude is due to the intrinsic loss of the metallic patch, and to the fact that the incident energy is split into both reflected and transmitted waves. The π-limited phase shift range is a direct consequence of the fact that this structure only exhibits an electric dipolar response, as we shall next demonstrate. For this purpose, we compute the corresponding effective susceptibilities using the relations (4.11), which, for an x-polarized excitation and the simple geometrical shape of the structure, conveniently reduce to (4.22a) for χ_{ee}^{xx}, and to (4.22d) for χ_{mm}^{yy}. These susceptibility functions are now obtained by inserting the simulated transmission and reflection coefficients in Figure 6.3 into (4.22a) and (4.22d) and are plotted in Figure 6.4.

---

5 Since the patch is elongated in the x-direction, it will interact more strongly when excited by an x-polarized wave than a y-polarized one. We thus restrict our attention to x-polarization.

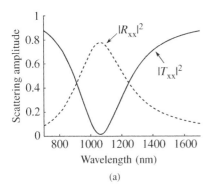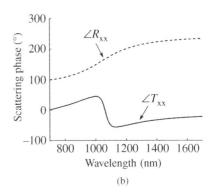

(a)   (b)

Figure 6.3 Reflection and transmission coefficients for the structure in Figure 6.2 with parameters $P = 300$ nm, $L = 200$ nm, $W = 80$ nm, and $H = 30$ nm. Since the particle is designed for the optical regime, the dielectric substrate is modeled as a half-space medium ($T_{\text{sub}} \to \infty$), as shown in Figure 6.1.

Figure 6.4 Real and imaginary parts of χ_{ee}^{xx} and χ_{mm}^{yy} for the structure in Figure 6.2 with scattering parameters in Figure 6.3. As mentioned in Section 2.2.2, the negative imaginary part of the susceptibilities corresponds to loss.

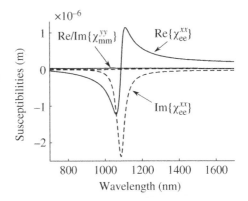

As can be seen, the electric susceptibility exhibits a typical Lorentzian resonance (see Section 2.2.2), while the magnetic response is negligible, as expected. We shall now demonstrate that a metasurface composed of scattering particles exhibiting only electric dipolar responses with Lorentzian profile has a maximal phase shift limited to π. Consider the metasurface reflection coefficient in (4.23b), which is directly obtained from the relations (4.22a) and (4.22d) used above. Setting $\chi_{mm}^{yy} = 0$ in that expression so as to remove the magnetic dipolar response and expressing the electric response with the Lorentzian function (2.23), transforms (4.23b) into

$$R_x = \frac{\omega \omega_p^2}{2jc_0 \left(\omega_0^2 - \omega^2\right) - \omega \omega_p^2}. \tag{6.1}$$

In this derivation, we have assumed that $\Gamma = 0$ in (2.23), which, although not consistent with the result shown in Figure 6.4, simplifies the expression (6.1) without

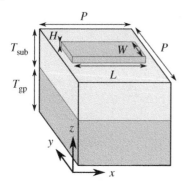

Figure 6.5 Simple metasurface scattering particle consisting of a thin metallic patch on top of a grounded dielectric substrate.

affecting our fundamental conclusion. The phase of R_x in (6.1) is

$$\text{Arg}(R_x) = \text{Arctan}\left[\frac{2c_0}{\omega\omega_p}\left(\omega_0^2 - \omega^2\right)\right]. \qquad (6.2)$$

Thus, when ω varies across the resonance frequency, ω_0, the phase of R_x undergoes a π phase shift due to the change of sign of the arctan function in (6.2). This result also holds if the magnetic dipolar response is nonzero and the electric one zero, and to the transmission coefficient for either dipolar responses.

We have established that a scattering particle with only an electric or only a magnetic dipolar response is not capable of fully controlling the phase of the waves scattered by a metasurface. We shall now investigate how to lift this restriction. Consider the scattering particle depicted in Figure 6.5, consisting of a metallic patch lying on a dielectric substrate backed by a ground plane.[6] Due to the presence of the ground plane, this structure does not allow any transmission and is thus limited to controlling the fields that are reflected by the metasurface.

To investigate the scattering response of the structure in Figure 6.5, we compute its scattering parameters and plot the resulting reflected amplitude and phase in Figure 6.6. This time, we also plot the parameters for the y-polarized wave for comparison with the x-polarized ones.

A resonant response is achieved for the x-polarized wave, whose resonant wavelength is redshifted compared to that in Figure 6.3 due to the presence of the ground plane. The y-polarized wave does not induce any resonance, as expected. The x-polarized wave is reflected with a fairly high reflection coefficient, which is limited only by ohmic losses, and exhibits the desired 0–2π phase coverage. We next compute the effective susceptibilities of this structure and plot them in Figure 6.7. We see that this structure induces both an electric dipolar response

---

6 The ground plane must be thicker than the skin depth of its material, which, in the optical regime, may require a substantial electric thickness. Note that the skin depth is defined as $\delta = \sqrt{2/(\omega\mu\sigma)}$ with σ being the conductivity of the material [81].

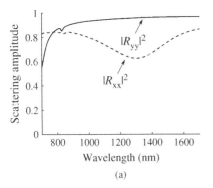
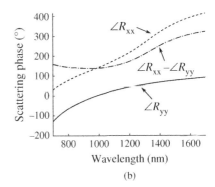

Figure 6.6 Reflection coefficients for the structure in Figure 6.5 with parameters $P = 300$ nm, $L = 200$ nm, $W = 80$ nm, $H = 30$ nm, $T_{sub} = 90$ nm, and $T_{gp} = 150$ nm. The scattering amplitudes are plotted in (a), while the corresponding phases are plotted in (b).

Figure 6.7 Real and imaginary parts of χ_{ee}^{xx} and χ_{mm}^{yy} for the structure in Figure 6.5.

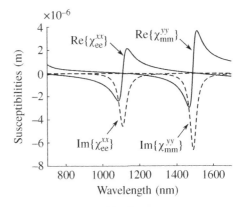

and a magnetic dipolar response, each individually providing up to π phase shift, which combines to a full 2π phase coverage.

To better understand why the structure in Figure 6.5 exhibits both electric and magnetic responses, while that in Figure 6.2 only exhibits an electric one, consider the schematic in Figure 6.8, which represents the cross-section of the patch in Figure 6.5 and its image according to image theory[7] [81]. Due to the symmetry of the structure (patch + image), the surface current, **J**, splits into two distinct modes: an even mode, where the current flows in the same direction in the patch and its image, corresponding to an electric dipole moment in the far-field, and an odd mode, where the current flows in opposite directions so as to form an equivalent current loop, corresponding to a magnetic dipole in the far-field.

---

7 Note that we neglect the presence of the dielectric spacer.

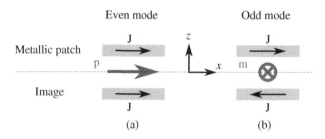

Figure 6.8 Mirror equivalent of the scattering particle in Figure 6.5 (dielectric spacer not considered). The surface current distribution, **J**, on both the image and the patch, can be either even, like in (a), or odd, like in (b), thus corresponding to electric or magnetic responses, respectively.

The fact that the scattering particle should have an electrically substantial spatial extent in the z-direction to provide a significant magnetic response is consistent with the discussion related to spatial dispersion in Section 2.3. Indeed, as apparent in (2.37), artificial magnetism appears as a consequence of (second-order) spatial dispersion, which can exist only if the scattering particles have a substantial electric thickness. Leveraging resonant effects, as in Figure 6.8, represents a simple but effective way to realize a strong magnetic response while keeping the scattering particle thickness in the subwavelength range.

A strategy to achieve full-phase coverage with a metallic scattering particle in transmission, instead of in reflection, is suggested in Figure 6.8. Indeed, if the scattering particle consisted of two metallic patches separated by a dielectric spacer instead of a metallic patch on top of a grounded dielectric slab, then the metasurface is applicable to transmission given the absence of ground plane. This strategy will be further detailed in Section 6.3.

Now that we have established that a metasurface with a scattering particle like the one in Figure 6.5 can achieve a full-phase coverage as function of the wavelength, as shown in Figure 6.6, we next explain how to achieve that full-phase coverage at the desired wavelength of operation. For this purpose, we create a look-up table by simulating the scattering particle at the wavelength of interest and varying only the length of its metallic patch. The resulting reflection amplitude and phase responses are plotted in Figure 6.9. As can be seen, the structure resonates at a length of about $L = 160$ nm around which the phase varies covering about 80% of the 0–2π phase range or 300° out of 360°, which corresponds to a phase gap of 60°.

According to Figure 6.9, this kind of structure suffers from the following two limitations: (i) Even though it exhibits the theoretical capability to provide full-phase coverage, it fails to realize this in practice because that would require a structure length much smaller than $L = 50$ nm, which, depending on the fabrication

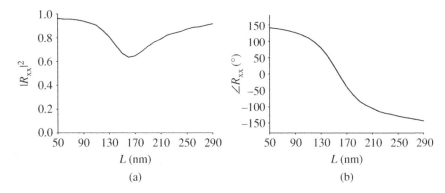

Figure 6.9 Reflection coefficient amplitude, in (a), and phase, in (b), for a patch length varying from $L = 50$ nm to $L = 290$ nm at a wavelength of $\lambda = 1064$ nm for the structure in Figure 6.5 with parameters $P = 300$ nm, $W = 80$ nm, $H = 30$ nm, $T_{sub} = 90$ nm, and $T_{gp} = 150$ nm.

technologies used, may be difficult to practically implement, and bigger than $L = 290$ nm, which in this case would exceed the size of the unit cell ($P = 300$ nm). Note that this limitation only affects metasurfaces that require 0 to 2π phase modulations to be discretized with more than 6 unit cells. Indeed, with a phase gap of 60°, the maximum number of unit cells within one full-phase modulation is $360/60 = 6$. (ii) This structure does not only induce a variation of phase, as the value of L changes, but also a variation of the amplitude, as shown in Figure 6.9a, which introduces errors.

6.2.1.2 Dielectric Scattering Particles

So far, we have only discussed scattering particles possessing metallic inclusions. However, there exists an entire class of scattering particles made only with dielectric material. We shall now discuss such dielectric particles.

The operation of an all-dielectric metasurface slightly differs from one based on metallic scattering particles. Although the general principle, i.e. controlling the electric and magnetic dipolar response of each unit cell, remains the same, this is accomplished in a different fashion. In the case of a metallic scattering particle, the fields penetrate little into the metallic inclusions and mostly induce surface currents. Consequently, at least two metallic layers are required to produce both electric and magnetic responses, as explained in Section 6.2.1.1. In contrast, in a dielectric scattering particle, the fields penetrate into the dielectric material, which forms a resonator by itself. Such a dielectric resonator exhibits TM and TE resonances, whose dominant modes may be associated with equivalent electric and magnetic dipolar responses [78, 104]. An all-dielectric metasurface can hence be formed by periodically arranging subwavelength dielectric resonators and its

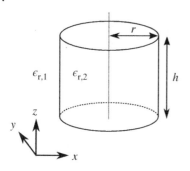

Figure 6.10 Example of a circular-cylindrical dielectric resonator. Due to its circular cross-section, this resonator exhibits an isotropic scattering response.

reflection and transmission coefficients can then be controlled by adjusting the shape of the resonators.

One of the most common shapes for a metasurface dielectric resonator is the circular cylinder [11, 82, 92], as the one depicted in Figure 6.10. Note that for mechanical support, the resonator is generally deposited on top of a transparent dielectric substrate such as glass, which is not shown in Figure 6.10. The response of the resonator essentially depends on the ratio between its relative permittivity and that of the surrounding medium, i.e. $\epsilon_{r,2}/\epsilon_{r,1}$. Since $\epsilon_{r,2}$ is much smaller for a dielectric than for a metal particle, the volume of the dielectric scatterer is larger than that of a generic metallic scattering particle. This leads to larger unit cell and may be an issue when implementing a metasurface requiring high spatial variations. However, dielectric structures are less lossy than their metallic counterparts, especially at optical frequencies, thus leading to better field transformation efficiencies. Moreover, due to their simple geometrical shape, dielectric resonators are often easier to fabricate.

The simple shape of dielectric resonators even allows closed-form expressions for their resonance frequencies. However, such expressions are not exact because the fields interacting with the resonator exist both inside and outside of it, which leads to a transcendental system of equations that cannot be solved analytically [118]. Nevertheless, for a dielectric resonator with circular cross section, as that in Figure 6.10, we may assume that $\mathbf{H}_{tan} \approx 0$ at the boundaries, if $\epsilon_{r,2} \gg 1$, which allows to derive approximate expressions of the resonance frequencies. For an *isolated* resonator standing in vacuum, for instance, the TE_{mnp} mode resonates at the wavelengths [118]

$$\lambda_{\text{TE},mnp} \approx \frac{2\pi\sqrt{\epsilon_{r,2}}}{\sqrt{\left(\frac{X_{mn}}{r}\right)^2 + \left(\frac{p\pi}{h}\right)^2}}, \tag{6.3}$$

where X_{mn} is the root of the Bessel function of order m, as

$$J_m(X_{mn}) = 0. \tag{6.4}$$

Figure 6.11 Electric and magnetic scattering cross-sections of a dielectric resonator with $r = 250$ nm, $h = 500$ nm and $\epsilon_{r,2} \approx 11.46$.

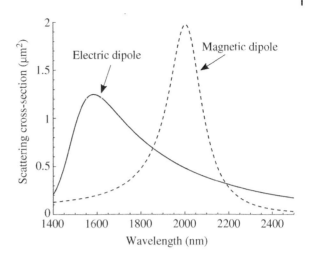

The first dominant mode is TE_{011}, for which $X_{mn} = X_{01} = 2.405$ [118]. Although expression (6.3) is an approximation, it still provides a good initial guess of the first resonance frequency of a dielectric resonator in terms of its size and material parameters.

To illustrate the validity of (6.3), consider the example of an *isolated* dielectric resonator made of amorphous silicon with $r = 250$ nm and $h = 500$ nm illuminated by an *x*-polarized plane wave with a desired resonant wavelength in the near- to short-infrared regime. We assume that the resonator is standing in vacuum without substrate for mechanical support and that $\epsilon_{r,2} \approx 11.46$. Substituting these parameters into (6.3) yields a resonance wavelength of $\lambda_{TE,001} \approx 1850$ nm. The electric and magnetic dipolar scattering cross-sections[8] of this resonator are plotted in Figure 6.11, where the first resonant mode, corresponding to a magnetic dipolar resonance, occurs at $\lambda = 2000$ nm and the next resonant mode, corresponding to an electric dipolar resonance, occurs around $\lambda = 1600$ nm. The former is only slightly redshifted compared to the value of $\lambda_{TE,mnp} \approx 1850$ nm predicted by (6.3). Figure 6.12 plots the electric fields on a *xz*-plane cross-section of the resonator. We easily recognize within the resonator the field distribution of an *x*-polarized dipole in Figure 6.12a, and the solenoidal electric field of a *y*-polarized magnetic dipole in Figure 6.12b.

Let us now investigate the scattering response of a metasurface formed by periodically repeating this resonator with a lattice period of 800 nm. The reflection and transmission amplitude and phase, still ignoring the presence of a substrate, are

8 The dipolar scattering cross-sections are obtained by first computing the electromagnetic scattering from an isolated dielectric resonator. Then, the corresponding dipolar contributions are calculated by multipolar decomposition, from which the dipolar scattering cross-sections may be computed [109].

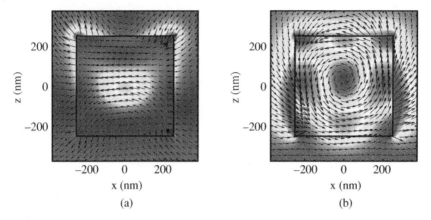

Figure 6.12 Cross-sectional views of the electric field distribution of the dielectric resonator in Figure 6.11. The field in (a) (*x*-oriented equivalent electric dipole) corresponds to a wavelength of 1520 nm, where the magnetic contribution is small compared to the electric one, while the field in (b) (*y*-oriented equivalent magnetic dipole) is at a wavelength of 2000 nm.

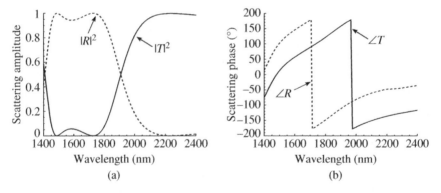

Figure 6.13 Reflection and transmission coefficients of a metasurface formed by the repetition of the circular-cylindrical dielectric resonator of Figure 6.11. (a) Amplitude. (b) De-embedded phase.

plotted in Figure 6.13. We see that with the chosen size parameters, the metasurface behaves as a reflector in the 1400–1900 nm range, where the reflection phase exhibits a full-phase coverage. Note that a transmission-type dielectric metasurface may be realized by changing the dimensions of the resonator so as to excite higher order TE and TM resonances, as will be shown in Section 7.2.

We now compute the metasurface electric and magnetic susceptibilities by inserting the scattering parameters shown in Figure 6.13 into (4.22a) and (4.22d). The results are plotted in Figure 6.14. They reveal that repeating the resonator

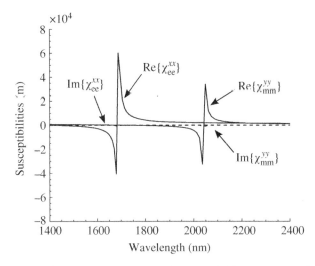

Figure 6.14 Electric and magnetic susceptibilities corresponding to the metasurface whose scattering parameters are plotted in Figure 6.13.

according to a subwavelength array results in a redshift of the susceptibility resonance wavelength compared to that of the isolated resonator in Figure 6.11. This redshift is a consequence of the capacitive coupling between adjacent resonators, which increases the electromagnetic density of the structure. Note that the presence of a substrate for mechanically supporting the resonators would result in an even more pronounced redshift due to the concentration of fields inside the substrate instead of vacuum. A look-up table, similar to that plotted in Figure 6.9, may be created, for instance, by fixing a desired wavelength of operation and varying the radius of the cylinder.

Finally, it is clear from symmetry that the circular-cylindrical shaped resonator in Figure 6.10 must lead to an isotropic metasurface. In order to realize, for instance, a birefringent metasurface, one would need to break the symmetry of the resonator in the xy-plane, which may be accomplished by an elliptical or rectangular cross-sectional shape instead of a circular one.

6.2.2 Circular Polarization

In Section 6.2.1, we have discussed how full-phase coverage is, in theory, achievable by changing the dimensions of metallic or dielectric scattering particles. However, this technique suffers from limited phase-shift coverage and spurious amplitude modulation, as shown in Section 6.2.1.1. We shall now present a solution to overcome these issues in the case of circularly polarized waves.

The method we shall present can provide a 100% full-phase coverage without amplitude modulation. It is based on the concept of the geometric phase, which is also referred to as Pancharatnam–Berry phase [18, 119]. It requires the incident wave to be circularly polarized and the scatterer to behave as a half-wave plate, i.e. a birefringent structure that induces a π phase shift between x- and y-polarized waves. If these two conditions are satisfied, the phase shift across the structure is controlled by a physical rotation of the scattering particle within the unit cell. This effect is best explained using Jones formalism [143].

Consider a half-wave plate that can be rotated by an arbitrary angle θ in its own plane, i.e. transverse to the direction of wave propagation in the case of normal incidence and that is illuminated by a circularly polarized wave. The polarization state of the transmitted[9] wave may be written as

$$\mathbf{J}_t = \overline{\overline{R}}(\theta) \cdot \overline{\overline{M}}_{\text{HWP}} \cdot \overline{\overline{R}}(-\theta) \cdot \mathbf{J}_i, \quad \text{and} \quad \overline{\overline{M}}_{\text{HWP}} = \begin{pmatrix} 1 & 0 \\ 0 & -1 \end{pmatrix}, \tag{6.5}$$

where $\mathbf{J}_i$ and $\mathbf{J}_t$ are the polarization states of the incident and transmitted waves, respectively, $\overline{\overline{M}}_{\text{HWP}}$ is the Jones matrix of a half-wave plate, and $\overline{\overline{R}}(\theta)$ is the rotation matrix defined as

$$\overline{\overline{R}}(\theta) = \begin{pmatrix} \cos\theta & -\sin\theta \\ \sin\theta & \cos\theta \end{pmatrix}. \tag{6.6}$$

Assuming that the incident wave is left-handed circularly polarized, Eq. (6.5) becomes

$$\mathbf{J}_t = \overline{\overline{R}}(\theta) \cdot \overline{\overline{M}}_{\text{HWP}} \cdot \overline{\overline{R}}(-\theta) \cdot \begin{pmatrix} 1 \\ i \end{pmatrix} = e^{j2\theta} \begin{pmatrix} 1 \\ -i \end{pmatrix}. \tag{6.7}$$

This shows that not only the transmitted wave polarization changes handedness but also, that, more importantly, it acquires a transmission phase shift equal to twice the rotation angle, θ. A similar effect occurs for a right-handed circularly polarized incident wave.

To implement a half-wave plate, one may use metallic scattering particles, as that in Figure 6.5, or dielectric particles such as that in Figure 6.10 but with an elliptical or a rectangular cross section, instead of a circular one, for birefringence. We shall next design a metallic half-wave plate using the reflective structure in Figure 6.5. It turns out that the scattering particle whose scattering parameters are plotted in Figure 6.6 already corresponds to a half-wave plate response. This can be seen by inspecting the phase difference between the x- and y-polarized reflected waves, plotted as the black line in Figure 6.6b. As can be seen, the phase difference is approximately equal to π within the wavelength range from 700 to 1200 nm. To

---

9 We consider the case of transmission for simplicity but similar considerations apply to the reflection case.

Figure 6.15 Co- and cross-polarized reflection coefficients for the structure in Figure 6.5 with parameters $P = 300$ nm, $L = 200$ nm, $W = 80$ nm, $H = 30$ nm, $T_{sub} = 90$ nm, and $T_{gp} = 150$ nm.

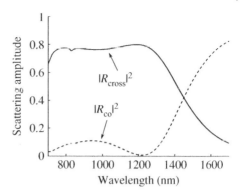

verify that the handedness of the reflected wave indeed changes and to characterize the reflection efficiency, we can compute the co- and cross-polarized reflection coefficients[10] using [143]

$$R_{co} = \frac{1}{2}\left[R_{xx} + R_{yy} - j(R_{xy} - R_{yx})\right], \tag{6.8a}$$

$$R_{cross} = \frac{1}{2}\left[R_{xx} - R_{yy} - j(R_{xy} + R_{yx})\right]. \tag{6.8b}$$

Since a half-wave plate changes the handedness of the light it scatters, it follows that an ideal reflective half-wave plate should have $R_{co} = 0$ and $R_{cross} = 1$. The resulting coefficients are plotted in Figure 6.15, where we have assumed that $R_{xy} = R_{yx} = 0$ because these cross terms are negligible for this particular structure.

The fact that the structure exhibits a phase shift difference between x- and y-polarized waves of π and that it essentially reflects a cross circularly polarized wave confirms that it indeed behaves as a half-wave plate with a reflection efficiency of about 80%. Now, controlling the phase of the reflected wave is simply achieved by rotating the rod within the unit cell plane to achieve the desired phase shift, as suggested in (6.7). Accordingly, the reflected wave complex amplitude, in terms of the metallic patch rotation angle θ, is

$$R = |R_{cross}|e^{2j\theta}. \tag{6.9}$$

For illustration, consider the design of a metasurface using the scattering particle in Figure 6.5 for an operation wavelength of 1200 nm, where $|R_{cross}| = 0.8$, as shown in Figure 6.15. The scattering response of this metasurface in terms of the metallic patch rotation angle, given by (6.9), is plotted in Figure 6.16. As expected, a variation of the rotation angle from 0 to π results in a linear increase of the reflection phase from 0 to 2π, while the reflection amplitude remains constant.

---

10 These relations may also be used with the transmission coefficients.

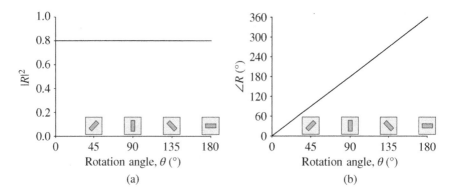

Figure 6.16 Reflection amplitude (a) and phase (b) in terms of the metallic patch rotation angle for the structure in Figure 6.5 when illuminated by a circularly polarized normally incident plane wave with a wavelength of 1200 nm. The insets correspond to a top view of the structure in Figure 6.5 and show the orientation of the metallic patch for different rotation angle.

6.3 Full-Phase Coverage with Perfect Matching

In Section 6.2, we have seen different strategies to achieve full-phase coverage, with linear and circular polarizations using metallic or dielectric scattering particles. However, the structures that were presented in Figures 6.2, 6.5, and 6.10 to realize these strategies suffer from some limitations. While the simplicity of their shape makes them particularly well suited for practically implementing metasurfaces, this simplicity also limits their field control capabilities, since they are essentially restricted to homoanisotropic metasurfaces, as evidenced by the retrieved susceptibilities. Moreover, when used in transmission metasurfaces, these structures cannot combine full-phase coverage *and* matching for each phase state because of their lack of degrees of freedom, which decreases the overall scattering efficiency of the metasurface.

In this section, we shall present a strategy that overcomes these limitations by enabling the implementation of transmission metasurfaces exhibiting full-phase coverage and perfect matching. Note that this strategy requires rather complex scattering particles that may be difficult to realize at optical frequencies, and is therefore generally better suited for the microwave regime.

Among all the possible candidates for microwave scattering particles, we shall next draw our attention to a particularly important class, which is characterized by the cascade of three metallic layers separated by dielectric spacers.[11] The rationale

---

11 Despite the number of layers, the overall structure remains deeply subwavelength. Note that this type of structures has also been realized in the infrared regime. However, in this range, it is not only challenging to fabricate, but it also strongly suffers from Ohmic losses.

Figure 6.17 Unit cell with subwavelength lateral dimensions P_x and P_y constituted of three cascaded metallic layers given in terms of their corresponding admittance Y_1, Y_2, and Y_3, and separated by dielectric spacers of thickness d, impedance η_d and propagation constant β.

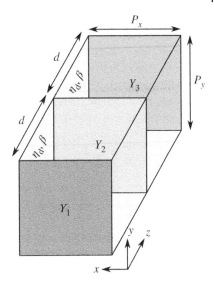

for using such kind of structures is that it corresponds to the minimum required number of layers, as will be shortly demonstrated, to achieve the specified requirements in terms of phase coverage and matching [74, 108, 111, 123, 124, 126, 127]. It also provides enough degrees of freedom to induce and control bianisotropic responses.

We next demonstrate that it is indeed possible to achieve full-phase coverage and matching by computing the scattering parameters of a generic three-metallic-layer structure. For simplicity, we assume that each layer is isotropic and consider the case of normally propagating waves. We apply the conventional ABCD matrix technique used in microwave theory [133]. A schematic of a generic unit cell is depicted in Figure 6.17, where three metallic layers, modeled by the uniform admittances Y_1, Y_2, and Y_3, are separated by two dielectric spacers of thickness d, impedance η_d, and propagation constant β.

For the case considered in Figure 6.17, the ABCD matrix of an admittance layer is [133]

$$\begin{pmatrix} 1 & 0 \\ Y_a & 1 \end{pmatrix}, \tag{6.10}$$

where $a = \{1, 2, 3\}$ and the ABCD matrix of a dielectric slab is

$$\begin{pmatrix} \cos(\beta d) & j\eta_d \sin(\beta d) \\ \frac{j\sin(\beta d)}{\eta_d} & \cos(\beta d) \end{pmatrix}, \tag{6.11}$$

where β is the propagation constant within the slab.

The general case $Y_1 \neq Y_2 \neq Y_3$ allows bianisotropic responses, thanks to the longitudinal asymmetry of the structure [4], as will be explained in Section 6.4.1. However, since we are interested in demonstrating the phase-coverage and matching properties of this type of unit cell, we shall next restrict our attention to the case of a *symmetric* particle, for which the two outer metallic layers are identical to each other, i.e. $Y_1 = Y_3$. The corresponding ABCD matrix of the entire symmetric unit cell is obtained, using (6.10) and (6.11), as

$$
\begin{pmatrix} A & B \\ C & D \end{pmatrix} = \begin{pmatrix} 1 & 0 \\ Y_1 & 1 \end{pmatrix} \cdot \begin{pmatrix} \cos(\beta d) & j\eta_d \sin(\beta d) \\ \frac{j\sin(\beta d)}{\eta_d} & \cos(\beta d) \end{pmatrix} \cdot \begin{pmatrix} 1 & 0 \\ Y_2 & 1 \end{pmatrix}
$$

$$
\cdot \begin{pmatrix} \cos(\beta d) & j\eta_d \sin(\beta d) \\ \frac{j\sin(\beta d)}{\eta_d} & \cos(\beta d) \end{pmatrix} \cdot \begin{pmatrix} 1 & 0 \\ Y_1 & 1 \end{pmatrix},
$$

(6.12)

The scattering parameters of this unit cell may be computed using the ABCD-to-scattering matrix conversion relation [133]

$$
\begin{pmatrix} S_{11} & S_{12} \\ S_{21} & S_{22} \end{pmatrix} = \frac{1}{2A + B/\eta_0 + C\eta_0} \begin{pmatrix} B/\eta_0 - C\eta_0 & 2 \\ 2 & B/\eta_0 - C\eta_0 \end{pmatrix}.
$$

(6.13)

We now investigate the scattering response of this structure by varying the values of Y_1 and Y_2. For this purpose, we substitute (6.12) into (6.13), and plot the amplitude and phase of the resulting transmission coefficient, S_{21}, in Figure 6.18a and b, respectively. We have assumed here that Y_1 and Y_2 are purely imaginary values, which corresponds to a gainless and lossless materials [133]. As can be seen, there exist values of Y_1 and Y_2 that do indeed result in full transmission and a full-phase coverage, as highlighted by the solid line in these figures.

Now that we have established that a three-metallic stack can achieve the desired scattering response, we shall explain how to design the metallic layers. There is obviously a plethora of shapes that may be considered for that purpose [110], but we will next restrict our attention to the commonly used cross potent, depicted in Figure 6.19.

The main advantage of cross potent is the possibility of controlling x- and y-polarized waves almost independently. Indeed, assuming that $\theta = 0°$, the parameters L_y, W_x, and B_x mostly affect y-polarized waves, while L_x, W_y, and B_y mostly affect x-polarized waves. Therefore, we can generally decompose the cross potent into two orthogonal "dog-bone" structures, where the parameters B and L may be used to control the capacitive coupling between adjacent unit cells, while W and, to a lesser extent, L, control its inductive response.

Another advantage of the cross potent is that the folded shape of the cross and the strong capacitive coupling between adjacent cells leads to a very small unit cell, with a typical lateral side of about $\lambda_0/5$. This advantage comes at the price that this structure is rather complex to optimize due to its large number of parameters and

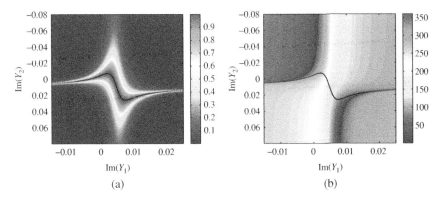

Figure 6.18 Scattering response of the unit cell in Figure 6.17 versus the purely imaginary values of Y_1 and Y_2 and with $Y_1 = Y_3$. (a) Transmitted power ($|S_{21}|^2$) and (b) phase of S_{21}. The solid line indicates that transmission phase values between 0 and 2π with full transmission can be achieved. The unit cell parameters are arbitrarily chosen as $\beta = 2\pi f \sqrt{\epsilon_r}/c_0$, $\eta_d = \eta_0/\sqrt{\epsilon_r}$, and $d = 1.5$ mm with $f = 10$ GHz and $\epsilon_r = 3$.

is more sensitive to a change of dimension due to the strong coupling between cells, which may be an issue for the design of nonuniform metasurfaces, but that may be mitigated by further full-wave optimization.

We now present the two general procedures for finding the dimensions of the crosses so as to achieve a given scattering response with a unit cell such as that in Figure 6.19b. For convenience, we will hereafter refer to the outer layer as layer 1 and the middle layer as layer 2. The first design method consists in designing each layer *individually*. For this purpose, we must first find the admittance value of each layer, which can be achieved by inserting the desired scattering parameters of the *entire unit cell* into (6.13), and solving for the corresponding ABCD parameters. The latter can then be inserted into (6.12), which can be solved for Y_1, Y_2, and Y_3. Although there is no direct method to relate the dimensions of the layers to their admittance value, we have simplified the task by reducing the total number of unknowns from the separate processing of each layer. The scattering parameters of each layer are then found by inserting $[A, B; C, D] = [1, 0; Y, 1]$, where Y is the admittance of a layer into (6.13). We then optimize the dimensions of the cross potent until Y matches the required admittance value of the layer. While this method may help simplifying the implementation of the unit cell, it should be noted that it usually leads to discrepancies between the scattering response of the layers taken separately or together because the ABCD method does not account for the evanescent coupling between the layers. This is not an issue if the layers are far apart from each other, but in metasurfaces, the layers are separated by dielectric spacers with typical thicknesses in the order of $\lambda_0/10$, implying that the coupling between the layers is not negligible. Therefore, this first method may be used to

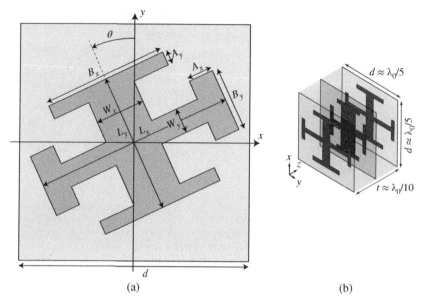

Figure 6.19 Representation of a symmetric unit cell with three metallic cross potent separated by dielectric slabs. (a) Top view of a generic cross potent with adjustable dimensions. (b) 3D perspective view.

obtain a good initial guess of the unit cell but definitively requires further optimization.

The second design method, which applies exclusively to symmetric cells ($Y_1 = Y_3$), consists in simulating the complete three-layer structure and optimizing all the dimensions at once. This may seem particularly tedious due to the large number of parameters involved but, as we will see, we may leverage the symmetry of the structure to simplify the task. As we have explained in Section 6.2, the subwavelength nature of the lateral size of the unit cell implies that only electric and magnetic dipolar modes are excited, and the control of the resonant property of these modes is therefore all that is required to control the scattering behavior of the metasurface. Due to the symmetry of the unit cell in its longitudinal direction ($Y_1 = Y_3$), we can apply an even/odd mode analysis similar to that used in Figure 6.8. The odd mode corresponds to a situation where the surface current distribution is equal and opposite on the two outer layers and null in the central one, hence resulting in a magnetic dipole moment. Since the current is null in the middle layer, the dimensions of its cross potent do not affect the magnetic response of the unit cell. On the other hand, the even mode corresponds to surface currents flowing in the same direction on all three layers, hence resulting in an electric dipole moment. This suggests that the unit cell may

Table 6.1 Geometrical dimensions (in mm) of a three-layer unit cell like the one in Figure 6.19 with $\theta = 0°$.

	L_x	L_y	W_x	W_y	A_x	A_y	B_x	B_y
Layer 1	4.25	4.75	0.625	0.25	0.5	0.5	4.25	3
Layer 2	3.75	5.25	0.5	0.375	0.5	0.5	2.25	4.5

be designed by first tuning the outer layers to control the magnetic response and then the middle layer to adjust the electric response. Moreover, considering that x and y polarizations can be treated essentially independently from each other, we can greatly reduce the number of parameters and optimize the response of the unit cell in a small number of iterations.

To illustrate this design procedure, we now investigate the effects of changing the physical dimensions of layers 1 and 2 on the unit-cell scattering response and corresponding susceptibilities. For this purpose, we consider a unit cell with dimensions listed in Table 6.1, which corresponds to a metasurface exhibiting full transmission at 10 GHz.

To see how changing layers 1 and 2 affects the scattering parameters of this unit cell, we perform three different simulations, which we next describe and whose results are reported in Figure 6.20.

The first simulation is performed with the exact same dimensions as those in Table 6.1, and the corresponding results are plotted in solid lines in the figures. For the second simulation, the value of L_x of layer 1 is increased from 4.25 to 5 mm, and the corresponding results are plotted in dotted lines in the figures. For the third simulation, the value of L_x of layer 2 is increased from 3.75 to 4.4 mm, and the corresponding results are plotted in dashed lines in the figures.

As explained above, we expect that these variations of dimensions will mostly affect the x-polarized transmission coefficient, S_{21}^{xx}, while leaving essentially unchanged the y-polarized transmission coefficient, S_{21}^{yy}. The amplitude of these scattering parameters is plotted in Figure 6.20a and b, respectively. We see that the scattering parameters of this unit cell have experienced very different variations. As expected, the parameter S_{21}^{xx} has drastically changed, while the parameter S_{21}^{yy} has remained almost unaffected, over the selected frequency range.

To investigate the effects of these dimensional changes on the electric and magnetic responses of the unit cell, we convert the scattering parameters in Figure 6.20a into the corresponding electric and magnetic susceptibilities using relations (4.22a) and (4.22d). The resulting real parts of χ_{ee}^{xx} for the three simulations are plotted in Figure 6.20c, while the real parts of χ_{mm}^{yy} are plotted in Figure 6.20d. We see that changing layer 1 has a strong effect on the magnetic

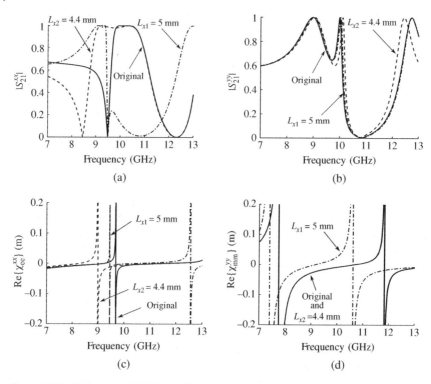

Figure 6.20 Full-wave simulations of three different designs for the unit cell in Figure 6.19. The solid curves correspond to the simulations of the original unit cell, whose dimensions are given in Table 6.1. The plots (a) and (b) correspond to the amplitude of S_{21}^{xx} and S_{21}^{yy}, respectively. The plots (c) and (d) are the real parts of χ_{ee}^{xx} and χ_{mm}^{yy}, respectively. For the dotted lines, L_x of layer 1 is changed to 5 mm. For the dashed lines, L_x of layer 2 is changed to 4.4 mm.

resonance and a non-negligible effect on the electric resonance (see the dotted lines). On the other hand, changing layer 2 only affects the electric resonance, as expected. In both cases, we have increased the dimensions L_x of the middle and outer layers, which translates in an increase of the capacitive coupling between adjacent unit cells and thus a decreases in the corresponding electric and/or magnetic resonance frequency, as visible by the frequency shift of the susceptibilities.

6.4 Effects of Symmetry Breaking

We have seen a few simple examples of scattering particles and their associated effective susceptibilities in Section 6.2. For such simple structures, it is generally simple to guess, even before performing any numerical simulations and using the scattering-to-susceptibility mapping relations (4.11), to which effective susceptibility tensor components they correspond to. For instance, it was straightforward to associate the x-oriented electric dipoles of the structures in Figure 6.2 to the effective susceptibility component χ_{ee}^{xx}. Similarly, the current loop that forms in the structure of Figure 6.5 was easily associated with the magnetic component χ_{mm}^{yy}. We thus see that, at least in some simple cases, we can intuitively connect the susceptibilities and the scattering particles' shape. However, such intuitive relationships are much harder to apply in the general case of bianisotropic scattering particles.

In order to facilitate the implementation of metasurfaces with features that are more complex than those in Figures 6.2 and 6.5, we will now discuss some general rules related to the symmetry of the particles that may be used to associate susceptibility components with the scattering response of a metasurface, even in the case of bianisotropy. Additionally, we will see in Section 6.4.1 that breaking the symmetry of a scattering particle in the longitudinal plane of the metasurface affects its angular scattering response and show how this may be related to the presence of normal polarizations. Then, in Section 6.4.2, we will see how breaking the symmetry of a scattering particle in the transverse plane of the metasurface affects the polarization state of the waves.

6.4.1 Angular Scattering

Understanding the relation between the angular scattering response of a metasurface and its susceptibilities is instrumental to the practical implementation of efficient metasurfaces. We will next demonstrate that the structural symmetries of a scattering particle in the longitudinal direction of the metasurface are intimately connected to the metasurface angular scattering response and hence to the presence of certain susceptibility components [5]. Note that we shall restrict our attention to metasurfaces that are spatially uniform and that do not affect the polarization of the waves.

For simplicity but without loss of generality, we consider the case of p-polarized ($E_y = H_x = H_z = 0$) waves propagating with respect to the xz-plane. We start by characterizing the angular scattering response of the metasurface by computing its reflection and transmission coefficients assuming an obliquely incident plane wave, propagating in the $\mp z$-directions, as shown in Figure 6.21. Here, each quadrant of the xz-plane is associated with a receiving and transceiving "port". Port 1 can communicate with port 2 by reflection (i.e. R_{21}), to port 3 by transmission (i.e. T_{31}) and cannot communicate to port 4 at all, and so on and so forth for the other ports.

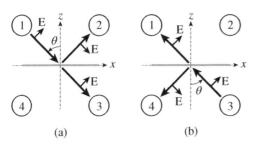

Figure 6.21 Angular scattering by a uniform metasurface that does not affect the polarization of the waves. Incidence angle and field polarization for (a) downward wave incidence and (b) an upward wave incidence.

(a)   (b)

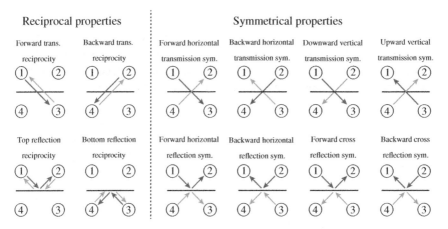

Figure 6.22 Representations of the 12 different scattering properties [5].

Using the convention in Figure 6.21, we can deduce all the possible reciprocity and angular scattering properties that apply to such a metasurface. These properties are presented in Figure 6.22, where each diagram represents an equality between two transmission or reflection coefficients. For instance, if the "forward transmission reciprocity" condition is satisfied, then the transmission coefficients T_{31} and T_{13} must be equal to each other. This figure shows the differences between reciprocal scattering and symmetrical scattering. Note that it may be easy to confuse asymmetric scattering with nonreciprocal scattering. However, these two concepts are very different from each other. Indeed, asymmetric scattering may be achieved by controlling the geometry of the scattering particles, while nonreciprocity can only be achieved through the presence of a time-odd external bias, as explained in Section 2.4.

For a reciprocal metasurface, a consequence of these properties is that if at least one of the transmission or reflection symmetry properties is satisfied, then all other transmission or reflection symmetry properties are also de facto satisfied. This is straightforward to demonstrate by writing down the equality relations for each case and combining them together.

To provide a more detailed description of these angular scattering properties, we next apply the GSTCs to compute the relationships between the angular reflection and transmission coefficients and the susceptibilities. We first define the differences and averages of the fields, for the two scenarios in Figure 6.21, as

$$\Delta \mathbf{E} = \pm \hat{\mathbf{x}} \frac{k_z}{k_0} (1 + R - T), \tag{6.14a}$$

$$\Delta \mathbf{H} = \hat{\mathbf{y}} \frac{1}{\eta_0} (-1 + R + T), \tag{6.14b}$$

$$E_{x,\text{av}} = \frac{k_z}{2k_0} (1 + T + R), \tag{6.14c}$$

$$E_{z,\text{av}} = \frac{k_x}{2k_0} (1 + T - R), \tag{6.14d}$$

$$H_{y,\text{av}} = \mp \frac{1}{2\eta_0} (1 + T - R), \tag{6.14e}$$

where $k_z = k_0 \cos \theta$ and $k_x = k_0 \sin \theta$, and where the top (bottom) sign corresponds to an incident wave propagating backward (forward) along the z-direction, as in Figure 6.21a (Figure 6.21b). Inserting these definitions into (4.2) and solving the resulting system for R and T, yields[12]

$$R = \frac{2}{C_b} \left\{ k_x^2 \chi_{ee}^{zz} - k_z^2 \chi_{ee}^{xx} - k_z \left[k_x (\chi_{ee}^{xz} - \chi_{ee}^{zx}) \mp k (\chi_{em}^{xy} - \chi_{me}^{yx}) \right] \right.$$
$$\left. \mp k k_x (\chi_{em}^{zy} + \chi_{me}^{yz}) + k^2 \chi_{mm}^{yy} \right\}, \tag{6.15a}$$

$$T = \frac{j k_z}{C_b} \left\{ k_x^2 (\chi_{ee}^{xz} \chi_{ee}^{zx} - \chi_{ee}^{xx} \chi_{ee}^{zz}) + (2j \mp k \chi_{em}^{xy})(2j \mp k \chi_{me}^{yx}) \right.$$
$$\left. + k_x \left[\chi_{ee}^{zx}(2j \mp k \chi_{em}^{xy}) + \chi_{ee}^{xz}(2j \mp k \chi_{me}^{yx}) \pm k \chi_{ee}^{xx}(\chi_{em}^{zy} + \chi_{me}^{yz}) \right] - k^2 \chi_{ee}^{xx} \chi_{mm}^{yy} \right\}, \tag{6.15b}$$

where the parameter C_b is given by

$$C_b = 2 \left[k_z^2 \chi_{ee}^{xx} + k_x^2 \chi_{ee}^{zz} \mp k k_x (\chi_{em}^{zy} + \chi_{me}^{yz}) + k^2 \chi_{mm}^{yy} \right] - j k_z \left[k_x^2 (\chi_{ee}^{xz} \chi_{ee}^{zx} - \chi_{ee}^{xx} \chi_{ee}^{zz}) \right.$$
$$\left. + 4 \pm k^2 (\chi_{ee}^{xx} \chi_{mm}^{yy} - \chi_{em}^{xy} \chi_{me}^{yx}) \mp k k_x (\chi_{ee}^{zx} \chi_{em}^{xy} + \chi_{ee}^{xz} \chi_{me}^{yx} - \chi_{ee}^{xx}(\chi_{em}^{zy} + \chi_{me}^{yz})) \right]. \tag{6.16}$$

We see that some susceptibility components are multiplied by the factors k_x, k_z, k_x^2, and/or k_z^2 implying that the presence of some of these susceptibility components affects the symmetrical angular scattering properties of the metasurface. In particular, the presence of the term k_x, related to $\sin \theta$, results in an asymmetric

---

12 We assume here the general case where the metasurface may be nonreciprocal.

scattering response for positive or negative values of θ, while k_z, k_x^2, and k_z^2 are related to symmetric angular responses with respect to θ.

We next consider the case of a reciprocal metasurface, which satisfies relations (2.52). We can deduce the following four important consequences from (6.15):

1) The presence of χ_{ee}^{xx}, χ_{mm}^{yy}, and χ_{ee}^{zz} does not break any angular scattering symmetry since these components are not associated with k_x.
2) In contrast, the components χ_{ee}^{xz} and χ_{ee}^{zx} are related to k_x but only affect the scattering symmetry of the transmission coefficients since, by reciprocity, the reflection coefficients must be angularly symmetric, i.e. $R_{12} = R_{21}$ and $R_{34} = R_{43}$.
3) The components χ_{em}^{xy} and χ_{me}^{yx}, although not directly related to k_x, are still associated with different reflection coefficients, via the $\pm$ signs in (6.15) and (6.16), which implies different reflection coefficients for forward and backward illuminations.
4) The scattering of a reciprocal metasurface is not affected by the presence of χ_{em}^{zy} and χ_{me}^{yz} since these terms cancel each other according to (2.52).

Now that we have established a connection between the angular scattering response of a metasurface and its susceptibilities, we shall investigate how this relates to the shape of the scattering particles and their structural symmetries [5]. For this purpose, we again restrict our attention to reciprocal metasurfaces. Referring to (6.15), we identify four distinct cases of angular scattering behavior that correspond to four different types of scattering particles, as we shall shortly demonstrate. These four cases are:

1) Symmetric transmission and reflection, i.e. $T(\theta) = T(-\theta)$ and $R(\theta)^{(+)} = R(\theta)^{(-)}$, where $(+)$ and $(-)$ denote forward and backward illumination, respectively.
2) Asymmetric transmission and symmetric reflection, i.e. $T(\theta) \neq T(-\theta)$ and $R(\theta)^{(+)} = R(\theta)^{(-)}$.
3) Symmetric transmission and asymmetric reflection, i.e. $T(\theta) = T(-\theta)$ and $R(\theta)^{(+)} \neq R(\theta)^{(-)}$.
4) Asymmetric transmission and reflection, i.e. $T(\theta) \neq T(-\theta)$ and $R(\theta)^{(+)} \neq R(\theta)^{(-)}$.

To investigate these four cases in terms of their relation with the type of scattering particles, we consider four different simple scattering particles made of subwavelength metallic rods, whose dimensions are detailed in [5]. For each particle, a series of simulations are performed, with periodic boundary conditions, vacuum as the background medium and a p-polarized incident wave, as in Figure 6.21, with an incidence angle that varies in the range $\theta = \pm 85°$ for both forward and backward illumination. The resulting scattering particles are depicted in

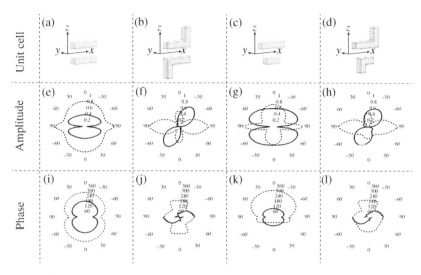

Figure 6.23 Angular scattering properties of four different reciprocal metasurfaces [5]. Top row, metasurface unit cells composed of two metallic rods vertically aligned in the *xz*-plane and periodically repeated in the *xy*-plane to form the metasurfaces. Middle row, amplitude of the transmission (solid lines) and reflection (dashed-dotted lines) coefficients versus incidence angle. Bottom row, phase of the transmission and reflection coefficients.

Figure 6.23a–d, and their angle-dependent reflection and transmission scattering amplitude and phase are plotted in Figure 6.23e–h and Figure 6.23i–l, respectively. Note that the plots in Figure 6.23e–l do not correspond to conventional radiation pattern: they represent the reflection and transmission coefficients, amplitude and phase, as functions of the incidence angle following the convention adopted in Figure 6.21a and b.

Finally, we associate the structural symmetries of the scattering particles in Figure 6.23 to the angular scattering symmetries of the corresponding metasurfaces and, using (6.15), to the susceptibilities involved. The resulting relationships are given in Table 6.2, where C_2 refers to a twofold (180°) rotation symmetry around the *y*-axis, while σ_{yz} and σ_{xy} refer to reflection symmetries through the *yz*-plane and the *xy*-plane, respectively. The table also provides the typical susceptibility components that would correspond to these different cases [5]. Note that these three symmetry properties (σ_{yx}, σ_{yz} and C_2) are connected to each other in such a way that if an element exhibits both C_2 and σ_{yz} symmetries, then it necessarily also exhibits a σ_{xy} symmetry due to the fundamental property that $C_2 \cdot \sigma_{yz} = \sigma_{xy}$ [32]. However, the opposite is not necessarily true, i.e. an element exhibiting a σ_{xy} symmetry does not necessarily exhibit both C_2 and σ_{yz} symmetries.

Table 6.2 Symmetry relationships between angular scattering and unit cell structure for the four types of reciprocal metasurfaces.

Structure	Reflection	Transmission	Type
$C_2\sigma_{yz}$ (or σ_{xy})	$C_2\sigma_{yz}$	$C_2\sigma_{yz}$	Birefringent $\chi_{ee}^{xx}, \chi_{mm}^{yy}$
C_2	$C_2\sigma_{yz}$	C_2	Anisotropic $\chi_{ee}^{xx}, \chi_{mm}^{yy}\ \chi_{ee}^{xz}, \chi_{ee}^{zx}, \chi_{ee}^{zz}$
σ_{yz}	σ_{yz}	$C_2\sigma_{yz}$	Bianisotropic $\chi_{ee}^{xx}, \chi_{mm}^{yy}\ \chi_{em}^{xy}, \chi_{me}^{yx}$
—	σ_{yz}	C_2	Bianisotropic $\chi_{ee}^{xx}, \chi_{mm}^{yy}, \chi_{ee}^{zz}\ \chi_{ee}^{xz}, \chi_{ee}^{zx}, \chi_{em}^{xy}, \chi_{me}^{yx}, \chi_{em}^{zy}, \chi_{me}^{yz}$

Table 6.2 shows that the angular reflection coefficients always exhibit a σ_{yz} symmetry, while the transmission coefficients always exhibit a C_2 symmetry. This is a consequence of reciprocity, which forces some of the scattering parameters to be equal to each other, as illustrated in Figure 6.22. We also see that the angular scattering behavior exhibits the same symmetries as the structural symmetries of the corresponding scattering particle (in addition to those due to reciprocity).

6.4.2 Polarization Conversion

We have just seen how the structural symmetries of a scattering particle in the transverse plane affect the angular scattering response of the metasurface. We shall now investigate how the shape of a scattering particle and its symmetries affect the polarization state of the waves scattered by a metasurface and how this relates to the metasurface susceptibilities.

To characterize the effects of an electromagnetic structure on the polarization state of the waves that it scatters, it is common practice to apply Jones formalism [73]. Assuming a transmission metasurface illuminated by a normally incident plane wave, the Jones matrix, which describes the polarization response of the metasurface, is simply the metasurface transmission matrix $\overline{\overline{S}}_{21}$, or $\overline{\overline{S}}_{12}$ depending on the direction of wave propagation, which can be obtained with (4.16).

To relate the metasurface susceptibilities to its effect on the state of polarization, we next assume a simplified scenario. Let us consider a reciprocal reflectionless bianisotropic gainless and lossless metasurface surrounded by vacuum. Inserting the reflectionless conditions (4.20) and the gainless and lossless conditions (2.77) into (4.16) and solving the resulting system for the components of $\overline{\overline{S}}_{21}$ yields[13]

$$A = \frac{k^2 \left((\chi_{ee}^{xy})^2 - \kappa^2 - \chi_{ee}^{xx}\chi_{ee}^{yy}\right) + 2jk(\chi_{ee}^{xx} - \chi_{ee}^{yy}) - 4}{k^2 \left(\kappa^2 + \chi_{ee}^{xx}\chi_{ee}^{yy} - (\chi_{ee}^{xy})^2\right) - 2jk(\chi_{ee}^{xx} + \chi_{ee}^{yy}) - 4}, \tag{6.17a}$$

$$B = \frac{4jk(\chi_{ee}^{xy} - \kappa)}{k^2 \left(\kappa^2 + \chi_{ee}^{xx}\chi_{ee}^{yy} - (\chi_{ee}^{xy})^2\right) - 2jk(\chi_{ee}^{xx} + \chi_{ee}^{yy}) - 4}, \tag{6.17b}$$

$$C = \frac{4jk(\kappa + \chi_{ee}^{xy})}{k^2 \left(\kappa^2 + \chi_{ee}^{xx}\chi_{ee}^{yy} - (\chi_{ee}^{xy})^2\right) - 2jk(\chi_{ee}^{xx} + \chi_{ee}^{yy}) - 4}, \tag{6.17c}$$

$$D = \frac{k^2 \left((\chi_{ee}^{xy})^2 - \kappa^2 - \chi_{ee}^{xx}\chi_{ee}^{yy}\right) - 2jk(\chi_{ee}^{xx} - \chi_{ee}^{yy}) - 4}{k^2 \left(\kappa^2 + \chi_{ee}^{xx}\chi_{ee}^{yy} - (\chi_{ee}^{xy})^2\right) - 2jk(\chi_{ee}^{xx} + \chi_{ee}^{yy}) - 4}, \tag{6.17d}$$

where $(A, B, C, D) = (S_{21}^{xx}, S_{21}^{xy}, S_{21}^{yx}, S_{21}^{yy})$ for compatibility with the Jones matrix convention,[14] and κ is a chirality parameter given by $\overline{\overline{\chi}}_{em} = -\overline{\overline{\chi}}_{me} = \kappa \overline{\overline{I}}$ (see (4.21)). Relations (6.17) provide a relationship between susceptibilities and incident and transmitted electric fields since

$$\begin{bmatrix} E_{t,x} \\ E_{t,y} \end{bmatrix} = \begin{bmatrix} A & B \\ C & D \end{bmatrix} \cdot \begin{bmatrix} E_{i,x} \\ E_{i,y} \end{bmatrix}, \tag{6.18}$$

where $\mathbf{E}_i = \mathbf{x}E_{i,x} + \hat{\mathbf{y}}E_{i,y}$ and $\mathbf{E}_t = \mathbf{x}E_{t,x} + \hat{\mathbf{y}}E_{t,y}$ are the incident and transmitted electric field vectors at the metasurface plane, respectively. The $ABCD$-matrix in (6.18) is thus the Jones matrix of the metasurface, whose susceptibilities may be expressed in terms of the Jones parameters by reversing (6.17), as

$$\chi_{ee}^{xx} = \frac{2j}{k_0}\left[\frac{A - BC - 1 + D(A - 1)}{A - BC + 1 + D(A + 1)}\right], \tag{6.19a}$$

$$\chi_{ee}^{yy} = -\frac{2j}{k_0}\left[\frac{A + BC + 1 - D(A + 1)}{A - BC + 1 + D(A + 1)}\right], \tag{6.19b}$$

$$\chi_{ee}^{xy} = \frac{2j}{k_0}\left[\frac{B + C}{A - BC + 1 + D(A + 1)}\right], \tag{6.19c}$$

$$\kappa = \frac{2j}{k_0}\left[\frac{B - C}{A - BC + 1 + D(A + 1)}\right]. \tag{6.19d}$$

We identify in (6.17) and (6.19) seven types of scattering responses, which we shall now individually discuss. For this purpose, we use the scattering particles proposed in [84, 103], where the structural symmetries are explicitly related to

---

13 We have used the counterpart of (4.20a), i.e., $\overline{\overline{\chi}}_{mm} = -\overline{\overline{J}} \cdot \overline{\overline{\chi}}_{ee} \cdot \overline{\overline{J}}$, to express the magnetic susceptibilities in terms of the electric ones.
14 These ABCD parameters should not be confused with those discussed in Section 6.3.

Case	Examples of possible structures	Possible symmetries	Jones matrix	Susceptibilities
a		σ_{yz} σ_{xz} σ_{xy} $C_{Nz}, N>2$	$\begin{pmatrix} A & 0 \\ 0 & A \end{pmatrix}$	$\chi_{ee}^{xx} = \chi_{ee}^{yy} \neq 0$ $\chi_{ee}^{xy} = 0$ $\kappa = 0$
b	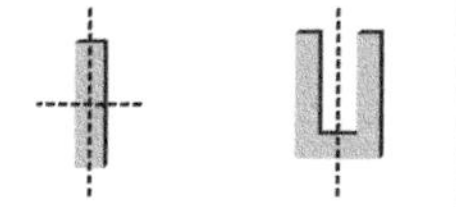	σ_{yz} σ_{xz} C_{2z} σ_{xy}	$\begin{pmatrix} A & 0 \\ 0 & D \end{pmatrix}$	$\chi_{ee}^{xx} \neq \chi_{ee}^{yy} \neq 0$ $\chi_{ee}^{xy} = 0$ $\kappa = 0$
c	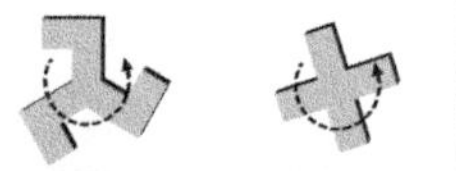	$C_{Nz}, N>2$	$\begin{pmatrix} A & B \\ -B & A \end{pmatrix}$	$\chi_{ee}^{xx} = \chi_{ee}^{yy} \neq 0$ $\chi_{ee}^{xy} = 0$ $\kappa \neq 0$
d	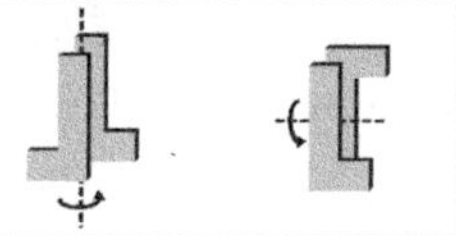	C_{2y} C_{2x}	$\begin{pmatrix} A & B \\ -B & D \end{pmatrix}$	$\chi_{ee}^{xx} \neq \chi_{ee}^{yy} \neq 0$ $\chi_{ee}^{xy} = 0$ $\kappa \neq 0$
e	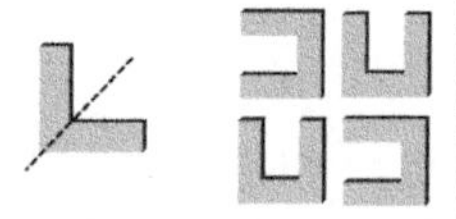	$\sigma_{xz, \pm 45°}$ σ_{xy}	$\begin{pmatrix} A & B \\ B & A \end{pmatrix}$	$\chi_{ee}^{xx} = \chi_{ee}^{yy} \neq 0$ $\chi_{ee}^{xy} \neq 0$ $\kappa = 0$
f	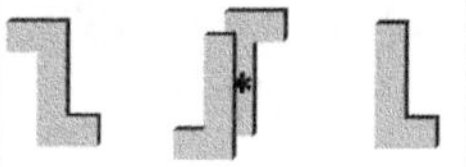	C_{2z} i σ_{xy}	$\begin{pmatrix} A & B \\ B & D \end{pmatrix}$	$\chi_{ee}^{xx} \neq \chi_{ee}^{yy} \neq 0$ $\chi_{ee}^{xy} \neq 0$ $\kappa = 0$
g		C_{2z}	$\begin{pmatrix} A & B \\ C & D \end{pmatrix}$	$\chi_{ee}^{xx} \neq \chi_{ee}^{yy} \neq 0$ $\chi_{ee}^{xy} \neq 0$ $\kappa \neq 0$

Figure 6.24 Relationships between scattering particle shapes and their symmetries, Jones matrix, and effective susceptibilities. These relationships apply to a metasurface made of a subwavelength periodic square lattice extending in the x- and y-directions. Note that the susceptibilities shown here are only indicative and some susceptibility components are missing, e.g. the z-oriented susceptibility components are not at all considered.

the Jones matrix of the corresponding metasurface. Using (6.19), we obtain their susceptibilities from the provided Jones matrix and, for each of the seven cases, present the scattering particle, structural symmetries, Jones matrix, and suscepti-bilities in Figure 6.24, where the scattering particles are represented as seen from a top-view above the xy-plane, as in Figure 6.19a. We assume that these structures are made of metallic material, as those discussed in Section 6.2. Note that in order to satisfy the reflectionless conditions (4.20), the structures in Figure 6.24 should be at least bi-layered (although most are not presented as such for convenience) in order to induce both electric and magnetic responses and hence cancel reflection, as explained in Section 6.2. We now discuss these seven types of structures individually.

Case a These very simple structures typically exhibit mirror symmetries along both the xz- and yz-planes (σ_{xz} and σ_{yz}) as well as rotation symmetry along the z-axis, C_{Nz} with $N > 2$. In terms of susceptibilities, they correspond to isotropic media where $\chi_{ee}^{xx} = \chi_{ee}^{yy}$. Therefore, the Jones matrix of the corresponding metasurface is diagonal, with identical x-to-x and y-to-y responses. Rotating a structure like the circular one within its unit cell by an arbitrary angle θ has no effect since

$$\bar{\bar{R}}(\theta) \cdot \begin{pmatrix} A & 0 \\ 0 & A \end{pmatrix} \cdot \bar{\bar{R}}(-\theta) = \begin{pmatrix} A & 0 \\ 0 & A \end{pmatrix}, \tag{6.20}$$

where $\bar{\bar{R}}(\theta)$ is the rotation matrix defined in (6.6). However, the square-shaped structure can only be rotated by multiples of 45° for the corresponding metasurface to still correspond to this Jones matrix. Indeed, rotating it by a different angle would lead to an overall metasurface with a lack of mirror symmetries along the xz- and yz-planes, thus leading to a more complicated effect on the polarization. As represented in the figure, these structures do not change the state of polarization of the scattered waves, and therefore $\chi_{ee}^{xy} = \kappa = 0$.

Case b Generalization of the structures of *textitCase a*, with different dimensions along x and y implying that $\chi_{ee}^{xx} \neq \chi_{ee}^{yy}$ (birefringence). They exhibit both or only one of the mirror symmetries along the xz- and yz-planes (σ_{xz} and σ_{yz}) and may exhibit a C_{2z} rotation symmetry along the z-axis. Their Jones matrix is diagonal with different x-to-x and y-to-y. They do not affect the polarization of x- or y-polarized waves. However, if the incident wave was diagonally polarized, or, equivalently, if the structure was rotated within the unit cell, they would behave as those of *Case f* since

$$\bar{\bar{R}}(\theta) \cdot \begin{pmatrix} A & 0 \\ 0 & D \end{pmatrix} \cdot \bar{\bar{R}}(-\theta)$$
$$= \begin{pmatrix} A\cos^2\theta + D\sin^2\theta & (A-D)\cos\theta\sin\theta \\ (A-D)\cos\theta\sin\theta & D\cos^2\theta + A\sin^2\theta \end{pmatrix} = \begin{pmatrix} A' & B' \\ B' & D' \end{pmatrix}, \tag{6.21}$$

which would result in polarization conversion effects. Note that in the special case where $\theta = \pi/4$, Eq. (6.21) reduces to

$$\overline{\overline{R}}(\pi/4) \cdot \begin{pmatrix} A & 0 \\ 0 & D \end{pmatrix} \cdot \overline{\overline{R}}(-\pi/4) = \frac{1}{2}\begin{pmatrix} A+D & A-D \\ A-D & A+D \end{pmatrix} = \begin{pmatrix} A'' & B'' \\ B'' & A'' \end{pmatrix},$$

(6.22)

which corresponds to the response of the structures of *Case e*.

Case c Structures exhibiting rotation symmetry so that C_{Nz} with $N > 2$ may be used to create chiral media. The first structure has a C_{3z} rotation symmetry and exhibits no σ_{xz} or σ_{yz} mirror symmetry, while the second one has a C_{4z} rotation symmetry as well as σ_{xz} and σ_{yz} (when the arms are aligned with the system of coordinates). On their own, these structures are not fundamentally chiral [27]. To create a chiral medium out of the first structure, it is enough to place it on top of a substrate, which breaks the symmetry of the system in the longitudinal direction resulting in an overall chiral response [27, 103, 130]. This strategy would not be sufficient for the second cross-shaped structure due to its additional mirror symmetries. However, it is still possible to create a chiral metasurface out of it by placing it on a substrate and rotating the cross within the unit cell. It should be rotated such that its arms are not aligned along the x and y axes or at $\pm45°$ from them. Assuming that the square lattice of the metasurface is on a xy grid, this rotation of the particle would effectively cancel the overall σ_{xz} and σ_{yz} mirror symmetries of the metasurface, hence making it chiral.

Breaking the longitudinal symmetry of the system may also be achieved by cascading several of these structures and changing their dimensions, composition, and orientation. This strategy has, for instance, been used to create a stronger chiral response [130].

This type of chiral media exhibit a chiral parameter $\kappa \neq 0$, while $\chi_{ee}^{xy} = 0$. These structures are the best for polarization rotation since their effect is rotation-invariant. Indeed, similar to *Case a*, we have that

$$\overline{\overline{R}}(\theta) \cdot \begin{pmatrix} A & B \\ -B & A \end{pmatrix} \cdot \overline{\overline{R}}(-\theta) = \begin{pmatrix} A & B \\ -B & A \end{pmatrix}.$$

(6.23)

Note that this is also the conclusion that was reached in Section 4.1.7.1 after synthesizing a chiral polarization rotating metasurface with the susceptibilities given by (4.46).

Case d These structures, which possess a C_{2x} or C_{2y} rotation symmetry, result in a generalized chiral response. They induce a counter-rotating effect on the fields, i.e. the metasurface Jones matrix is the negative of its own transpose, like the chiral structures of *Case c* but have different diagonal components ($\chi_{ee}^{xx} \neq \chi_{ee}^{yy}$) due to their unequal lengths in the x- and y-directions. Because of that, their effect on the polarization state of the scattered waves depends on the angular orientation of the metasurface, as shown in (6.21).

Case e Structures that have a ±45° mirror symmetry with respect to the *xz*- or *yz*-plane exhibit a nonzero *x*-to-*y* and *y*-to-*x* coupling, leading to polarization conversion but not chirality. As explained in *Case b*, such response can be obtained by rotating a birefringent scatterer within the unit cell by an angle of ±45°. Note that such structures also typically exhibit a mirror symmetry along the *xy*-plane, as those of *Case a* and *Case b*.

The unit cell composed of four split-ring resonators does not, per se, exhibit a ±45° mirror symmetry with respect to the *xz*- or *yz*-plane. However, in a metasurface composed of a periodic repetition of this unit cell, one can show that by reflecting the structure diagonally and shifting it by half a period along *x* or *y*, one can retrieve the original unit cell [35, 103], which implies that this structure still exhibits the type of response as the L-shaped one.

Case f These structures are a generalization of the structures of *Case e* but with different *x*-to-*x* and *y*-to-*y* responses due to their unequal lengths in the *x*- and *y*-directions. They can either exhibit a mirror symmetry with respect to the *xy*-plane, as the first and third depicted structures, or an inversion symmetry (*i*), as the second one. The same type of response can also be achieved by rotating a birefringent structure by a given angle, as explained in *Case b*.

Case g These structures present either no symmetry or possibly a C_{2z} rotational symmetry. They can be used to perform any operation on the wave polarization state providing that they do not violate the imposed conditions of reciprocity and energy conservation.

7

Applications

We have seen in Chapter 4 how to synthesize a metasurface in terms of its susceptibilities for specified electromagnetic transformations. In Chapter 5, we have seen how to analyze the scattering response of such a metasurface so as to both verify that it satisfies the specifications and investigate investigate how it scatters under arbitrary unspecified illumination conditions. Finally, the implementation of metasurface scattering particles has been discussed in Chapter 6. Thus, we are now fully equipped to present metasurface applications and detail their design from initial specifications in terms of field transformation to the final shape of their scattering particles.

Note that the purpose of this chapter is not to provide a complete review of all metasurface applications that have been reported to date. In fact, the vast majority of these applications revolve around rather simple working principles that we have already discussed in previous chapters. Indeed, most metasurface applications are related either to polarization transformations, as illustrated in Section 4.1.7.1 for the case of polarization rotation, and wave-propagation manipulations such as wave focalization/collimation, as already addressed in Sections 5.1 and 5.2.

We shall thus present some more advanced metasurface applications that are not trivial to implement. Accordingly, we shall discuss the following three applications: (i) angle-independent beam splitting, illustrated in Section 7.1 with a PEC/PMC angle-independent reflective metasurface; (ii) matching, illustrated in Section 7.2 with an anti-reflection metasurface; and (iii) fully efficient scattering, illustrated in Section 7.3 with a fully efficient refractive metasurface.

7.1 Angle-Independent Transformation

A metasurface may be required to operate not just for one specific angle of incidence, but over a large spectrum of angles. While, this specification may a priori seem to conflict with the fact that the angular scattering response of a metasurface generally depends on the angle of incidence, as explained in Sections 4.1.7.3 and 6.4.1, we shall see that, under specific conditions, angle-independent scattering is indeed possible.

To demonstrate this, consider Eq. (6.15), which provides the reflection and transmission coefficients of a bianisotropic metasurface that does not change the polarization of the waves and that includes normal susceptibility components. We assume that the metasurface is surrounded by the same medium at both of its sides, which is a necessary condition for angle-independent scattering.[1] Since the angular dependence is due to the presence of the parameters k_x and k_z, as evidenced in (6.15), eliminating these parameters would naturally suppress the angular dependence. This may be achieved by setting the homoanisotropic χ_{ee}^{uv} and χ_{mm}^{uv} susceptibility components to zero and keeping only the tangential heteroanisotropic components χ_{em}^{xy} and χ_{me}^{yx}. This indeed reduces (6.15) to

$$R = \pm \frac{2jk_0(\chi_{em}^{xy} - \chi_{me}^{yx})}{4 + k_0^2 \chi_{em}^{xy} \chi_{me}^{yx}}, \tag{7.1a}$$

$$T = \frac{(2 \pm jk_0 \chi_{em}^{xy})(2 \pm jk_0 \chi_{me}^{yx})}{4 + k_0^2 \chi_{em}^{xy} \chi_{me}^{yx}}, \tag{7.1b}$$

where the $\pm$ sign in (7.1a) corresponds to incident wave propagation in the $\pm z$-direction, respectively. As can be seen, these relations do not include any angular parameters. Thus, a tangential heteroanisotropic metasurface has angle-independent scattering properties. In the case of reciprocity, we further have $\chi_{me}^{yx} = -\chi_{em}^{xy}$ from (2.52), which simplifies (7.1) into

$$R = \pm \frac{4jk_0 \chi_{em}^{xy}}{4 - (k_0 \chi_{em}^{xy})^2}, \tag{7.2a}$$

$$T = \frac{4 + (k_0 \chi_{em}^{xy})^2}{4 - (k_0 \chi_{em}^{xy})^2}. \tag{7.2b}$$

If, in addition to reciprocity, we also require the absence of gain or loss, then χ_{em}^{xy} must be purely imaginary according to Table 2.2, which means that the imaginary part of χ_{em}^{xy} fully controls the scattering response of the metasurface. For illustration, the scattering parameters in (7.2) are plotted in Figure 7.1 versus $\mathrm{Im}\{\chi_{em}^{xy}\}/k_0$, which shows the angle-independent splitting between the reflected and transmitted waves.

At this point, one may wonder whether or not it may be possible to realize a metasurface such that $\chi_{ee} = \chi_{mm} = 0$ and $\chi_{em/me} \neq 0$. From a theoretical perspective, it is in fact impossible to have $\chi_{ee} = \chi_{mm} = 0$, since all structures

---

1 The presence of different media on both sides of a metasurface would lead to refraction by phase-matching and thus angle-dependent reflection and transmission coefficients.

Figure 7.1 Reflection and transmission coefficients (7.2) in terms of the metasurface heteroanisotropic susceptibility χ_{em}^{xy}.

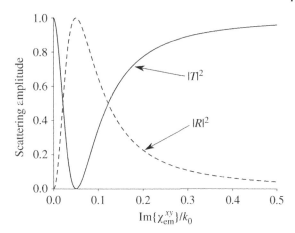

always exhibit at least a nonzero electric response. However, leveraging the resonant properties of the scattering particles, it is at least possible to achieve $\mathrm{Re}\{\chi_{ee}\} = \mathrm{Re}\{\chi_{mm}\} = 0$ at a given wavelength or frequency since the real part of these susceptibility functions follows a Lorentzian profile that changes signs at resonance, as shown in Figure 2.2. However, this would necessarily be associated with a nonzero imaginary part because of causality, as explained in Section 2.2.1. Nevertheless, it is still possible to assume that $\mathrm{Im}\{\chi_{ee}\} \approx 0$ and $\mathrm{Im}\{\chi_{mm}\} \approx 0$ by using very low loss materials.

According to the discussion in Section 6.4.1, the proposed metasurface must be asymmetric in the longitudinal direction since it is bianisotropic. Moreover, it should be spatially uniform since it is specified to leave the direction of wave propagation unchanged. This implies that we only need to design a single scattering particle and periodically repeat it to form the metasurface. We next assume a transverse magnetic (TM)-polarized incident wave at a frequency of about 10 GHz and use a simplified version of the cross potent shown in Figure 6.19. Specifically, we use a dog-bone shaped structure instead of a cross, for simplicity, since we are only interested in TM-polarized waves. Moreover, we do not need to have the reflection phase spanning the full phase range from 0 to 2π, so that we can only use two instead of three metallic layers.[2] We use the simulation setup described in Section 6.1 to design the scattering particle. The metasurface susceptibilities are obtained by solving 4.11 with a normally incident TM-polarized illumination. The dimensions of the scattering particle are then optimized so that $\mathrm{Re}\{\chi_{ee}^{xx}\} = \mathrm{Re}\{\chi_{mm}^{yy}\} = 0$ and $\mathrm{Im}\{\chi_{em}^{xy}\} = -\mathrm{Im}\{\chi_{me}^{yx}\} = 2/k_0$ near 10 GHz, which corresponds to the susceptibilities of the reflection maxima in Figure 7.1, as can be verified by

---

2 A single layer being unable to induce the required spatial asymmetry for the metasurface to be bianisotropic, as explained in Section 6.4.1.

Table 7.1 Geometrical dimensions (in mm) of the optimized two-layer dog-bone unit cell based on the cross potent in Figure 6.19 with $\theta = 0°$, $d = 6$ mm and $t = 1.52$ mm.

	L_x	W_y	A_x	B_y
Layer 1	5	0.5	0.5	4.5
Layer 2	5	0.5	0.5	2.5

The substrate is a Rogers RO3003 with $\epsilon_r = 3$.

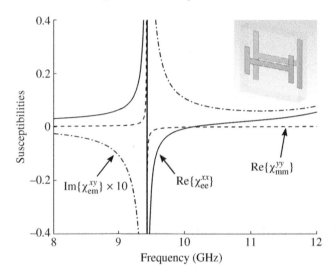

Figure 7.2 Susceptibilities of a tangential homoanisotropic metasurface consisting of a periodic arrangement of the unit cell shown in the inset. The desired condition $\chi_{ee} = \chi_{mm} = 0$ and $\chi_{em/me} \neq 0$ is obtained at 10 GHz.

substituting these values into (7.2), which yields $R = \mp 1$ and $T = 0$. The resulting optimized dimensions of the unit cell are given in Table 7.1.

The fact that $R = \mp 1$ means that this metasurface behaves as a perfect electric conductor (PEC) when illuminated in the $+z$-direction and as a perfect magnetic conductor (PMC) when illuminated in the $-z$-direction. The designed scattering particle is shown in the inset of Figure 7.2, while its susceptibilities are plotted in the figure.

The angular response of the metasurface is plotted in Figure 7.3 versus the frequency and for incidence angles ranging from $\theta = -80°$ to $\theta = +80°$ by steps of 10°. This shows that it is indeed possible to realize a metasurface that exhibits an angular-independent scattering response for a very large range of incidence angles.

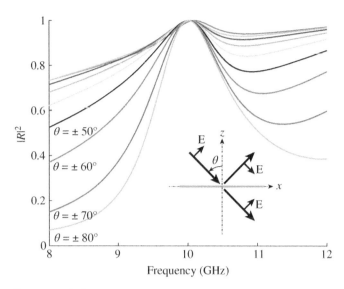

Figure 7.3 Reflectance of the metasurface in Figure 7.2 versus frequency and incidence angle obtained by full-wave simulation.

7.2 Perfect Matching

Being able to suppress reflection is of paramount importance to maximize the efficiency of transmission metasurfaces. In Section 4.1.4, we have seen that it possible to completely cancel the reflection by using tuned electric and magnetic dipolar responses. However, this was limited to the case of a metasurface surrounded by the same medium at both of its sides. The problem of canceling the reflection of a metasurface surrounded by two different media is more complicated and we shall address it now.

The schematic of Figure 7.4 depicts the problem under consideration. Consider a wave normally incident from medium 1 and transmitted into medium 2 via a reflection-canceling metasurface.[3] For practical reasons, we specifically aim at designing a passive, lossless, and reciprocal metasurface.

To synthesize such an antireflection metasurface, one may think that it is sufficient to specify the fields of the waves depicted in Figure 7.4 and solve the generalized sheet transition conditions (GSTC) relations (4.2). This procedure would indeed lead to a set of susceptibilities accomplishing this role but these susceptibilities would combine gain and loss, which is inefficient, unpractical, and most likely undesired. We shall therefore describe an alternative approach.

---

3 By reciprocity, reflection would also be canceled if the incident wave were coming from medium 2.

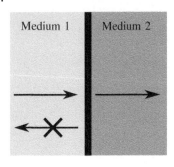

| Medium 1 | Medium 2 |

Figure 7.4 Metasurface operating as an anti-reflection coating for normally propagating waves between two different media.

When the surface impedance of the incident wave is different for incidence from the left and for the right sides, the metasurface may be synthesized twice, once for the problem when the incident wave propagates from the left to the right, and once for the reverse problem when the incident wave propagates from the right to the left. Intuitively, this may be understood as a strategy that provides the necessary extra degrees of freedom to perform the desired transformation, which, in this case, would lead to a metasurface without gain and loss.

Let us apply this concept to the problem of Figure 7.4. The surface impedance is defined as

$$Z_s = \frac{E_\parallel}{H_\parallel}, \tag{7.3}$$

where $E_\parallel$ and $H_\parallel$ are the electric and magnetic fields tangential to the metasurface, respectively. Therefore, the surface impedance is $Z_s = \eta_1$ for normal incidence from medium 1 and $Z_s = \eta_2$ for normal incidence from medium 2, where η_1 and η_2 are the intrinsic impedance of medium 1 and 2, respectively. Since these impedances are different, we may now proceed to the double synthesis assuming for simplicity that the waves are all x-polarized. For propagation from left to right, as in Figure 7.4, the fields in the plane of the metasurface are

$$E_{i,x} = 1, H_{i,y} = \frac{1}{\eta_1} \quad \text{and} \quad E_{t,x} = e^{j\alpha}\sqrt{\frac{\eta_2}{\eta_1}}, H_{t,y} = \frac{e^{j\alpha}}{\eta_2}\sqrt{\frac{\eta_2}{\eta_1}}, \tag{7.4}$$

where $\alpha \in \mathbb{R}$ is an arbitrary phase shift, the ratio $\sqrt{\frac{\eta_2}{\eta_1}}$ is required to satisfy energy conservation,[4] as specified in (2.81), and the reflected fields are assumed to be zero. The reverse problem is specified as

$$E_{i,x} = 1, H_{i,y} = -\frac{1}{\eta_2} \quad \text{and} \quad E_{t,x} = e^{j\alpha}\sqrt{\frac{\eta_1}{\eta_2}}, H_{t,y} = -\frac{e^{j\alpha}}{\eta_1}\sqrt{\frac{\eta_1}{\eta_2}}. \tag{7.5}$$

To obtain the metasurface susceptibilities, we now consider the bianisotropic GSTC relations in 4.3. Since polarization rotation is not required, we set to zero

---

4 We assume that η_1 and η_2 are purely real.

the off-diagonal components of $\overline{\overline{\chi}}_{ee}$ and $\overline{\overline{\chi}}_{mm}$ and the diagonal components of $\overline{\overline{\chi}}_{em}$ and $\overline{\overline{\chi}}_{me}$. The introduction of nonzero magnetoelectric coupling coefficients doubles the number of available susceptibility parameters. Therefore, we use here the multiple wave transformation technique described in Section 4.1.3, which yields the fully determined system of equations

$$
\begin{pmatrix} \Delta H_{y1} & \Delta H_{y2} \\ \Delta H_{x1} & \Delta H_{x2} \\ \Delta E_{y1} & \Delta E_{y2} \\ \Delta E_{x1} & \Delta E_{x2} \end{pmatrix} = \begin{pmatrix} \tilde{\chi}_{ee}^{xx} & 0 & 0 & \tilde{\chi}_{em}^{xy} \\ 0 & \tilde{\chi}_{ee}^{yy} & \tilde{\chi}_{em}^{yx} & 0 \\ 0 & \tilde{\chi}_{me}^{xy} & \tilde{\chi}_{mm}^{xx} & 0 \\ \tilde{\chi}_{me}^{yx} & 0 & 0 & \tilde{\chi}_{mm}^{yy} \end{pmatrix} \cdot \begin{pmatrix} E_{x1,av} & E_{x2,av} \\ E_{y1,av} & E_{y2,av} \\ H_{x1,av} & H_{x2,av} \\ H_{y1,av} & H_{y2,av} \end{pmatrix},
\tag{7.6}
$$

where the subscripts 1 and 2 refer to the fields in (7.4) and (7.5), respectively. Since we are restricting to x-polarized waves, the system (7.6) reduces to

$$
\begin{pmatrix} \Delta H_{y1} & \Delta H_{y2} \\ \Delta E_{x1} & \Delta E_{x2} \end{pmatrix} = \begin{pmatrix} \tilde{\chi}_{ee}^{xx} & \tilde{\chi}_{em}^{xy} \\ \tilde{\chi}_{me}^{yx} & \tilde{\chi}_{mm}^{yy} \end{pmatrix} \cdot \begin{pmatrix} E_{x1,av} & E_{x2,av} \\ H_{y1,av} & H_{y2,av} \end{pmatrix}.
\tag{7.7}
$$

We now substitute (7.4) and (7.5) into (7.7), and solve for the susceptibilities. We obtain, with (4.4),

$$
\chi_{ee}^{xx} = -\frac{4\eta_0 \sin \alpha}{2k_0 \sqrt{\eta_1 \eta_2} + k_0(\eta_1 + \eta_2)\cos \alpha},
\tag{7.8a}
$$

$$
\chi_{mm}^{yy} = -\frac{4\eta_1 \eta_2 \sin \alpha}{\eta_0[2k_0 \sqrt{\eta_1 \eta_2} + k_0(\eta_1 + \eta_2)\cos \alpha]},
\tag{7.8b}
$$

$$
\chi_{em}^{xy} = -\chi_{me}^{yx} = \frac{2j(\eta_1 - \eta_2)\cos \alpha}{2k_0 \sqrt{\eta_1 \eta_2} + k_0(\eta_1 + \eta_2)\cos \alpha},
\tag{7.8c}
$$

where η_0 and k_0 are the impedance and wavenumber in vacuum, respectively. Referring to Table 2.2, we see that these susceptibilities do indeed correspond to a gainless, lossless, and reciprocal metasurface. It is interesting to note that this solution requires the presence of bianisotropic susceptibility components, which appeared as a result of the specification of opposite wave incidence.

It may be practically desirable to avoid bianisotropy as this generally implies realization complexity. This can be done by setting $\alpha = \pm\pi/2$ in (7.8), which forces the fields of the transmitted waves in (7.4) and (7.5) to be imaginary and reduces (7.8) to

$$
\chi_{ee}^{xx} = \mp \frac{2}{k_0} \frac{\eta_0}{\sqrt{\eta_1 \eta_2}},
\tag{7.9a}
$$

$$
\chi_{mm}^{yy} = \mp \frac{2}{k_0} \frac{\sqrt{\eta_1 \eta_2}}{\eta_0},
\tag{7.9b}
$$

$$
\chi_{em}^{xy} = \chi_{me}^{yx} = 0.
\tag{7.9c}
$$

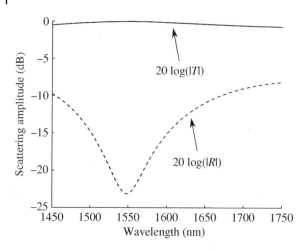

Figure 7.5 Reflection and transmission power coefficients of the antireflection metasurface showing a less than −20 dB matching at 1550 nm.

In practice, the phase of the transmitted waves is typically unimportant. Therefore, setting $\alpha = \pm\pi/2$ does not represent a restrictive condition.

We next provide an implementation example of the antireflection metasurface with susceptibilities (7.9). Referring to Figure 7.4, we select the relative permittivity of medium 1 and 2 to be $\epsilon_{r,1} = 1$ and $\epsilon_{r,2} = 6.25$, respectively. The metasurface is designed using the concept of dielectric resonators discussed in Section 6.2.1.2. After optimization for an operating wavelength of 1550 nm, the array periods are $P_x = P_y = 400$ nm and the resonator radius and height are $r = 150$ nm and $H = 1060$ nm with $\epsilon_r = 11.46$. The resulting scattering parameters are plotted in dB scale in Figure 7.5.

As can be seen, the metasurface achieves the excellent matching of $|T|^2 = 99.5\%$ at the specified wavelength. By comparison, the transmitted power through the system in the absence of a metasurface would be of

$$|T|^2 = 1 - \left| \frac{\sqrt{6.25} - 1}{\sqrt{6.25} + 1} \right|^2 = 81.6\%. \tag{7.10}$$

7.3 Generalized Refraction

Perfect refraction refers to the metasurface's capability to refract an incident wave with 100% transmission efficiency irrespective of the specified incidence and refraction angles. While refraction is usually a rather simple operation that only requires the generation of an appropriate phase gradient, achieving a 100% refraction efficiency in transmission is not trivial, as we will see next. Before providing the optimal solution, we shall first discuss why typical refractive metasurface synthesis methods are inefficient or limited.

7.3.1 Limitations of Conventional Synthesis Methods

As shown in Chapter 4, there are two main methods to synthesize a metasurface [7]. One consists in specifying the incident, reflected, and transmitted fields (without considering the reverse problem as in Section 7.2), while the other consists in specifying the scattering parameters. In order to highlight the differences and limitations of these two methods, we next consider the example of a refractive metasurface.

We assume that the metasurface is in vacuum, that the waves are TM-polarized and that refraction occurs in the xz-plane. The susceptibilities for the first method are given by (4.7) as

$$\chi_{ee}^{xx} = \frac{-\Delta H_y}{j\omega\epsilon_0 E_{x,av}}, \tag{7.11a}$$

$$\chi_{mm}^{yy} = \frac{-\Delta E_x}{j\omega\mu_0 H_{y,av}}, \tag{7.11b}$$

while those for the second method are given by (4.22) as

$$\chi_{ee}^{xx} = \frac{2j\left(T_x + R_x - 1\right)}{k_0\left(T_x + R_x + 1\right)}, \tag{7.12a}$$

$$\chi_{mm}^{yy} = \frac{2j\left(T_x - R_x - 1\right)}{k_0\left(T_x - R_x + 1\right)}. \tag{7.12b}$$

Note that the susceptibilities (7.12) are rigorously valid only for normally propagating waves, as explained in Section 4.1.4. However, as mentioned in Section 5.1, they may nonetheless be used as paraxial approximations.

The susceptibilities in (7.11) are synthesized for the fields[5]

$$\mathbf{E}_a = A_a \left(\hat{\mathbf{x}}\frac{k_{z,a}}{k_0} - \hat{\mathbf{z}}\frac{k_{x,a}}{k_0}\right) e^{-jk_{x,a}x} \quad \text{and} \quad \mathbf{H}_a = \frac{A_a}{\eta_0}\hat{\mathbf{y}}e^{-jk_{x,a}x}, \tag{7.13}$$

where a = i, t, r denotes the incident, transmitted, and reflected waves, respectively; A is their amplitude; and η_0 is the intrinsic impedance of the surrounding medium (considered here to be vacuum) and associated with the wavenumber k_0. The transverse and longitudinal wavenumbers are defined by $k_{x,a} = k_0 \sin\theta_a$ and

---

5 Refer to Figure 7.9a.

$k_{z,a} = k_0 \cos\theta_a$, respectively. Substituting the fields (7.13) into (7.11) and assuming that $A_r = 0$ for maximum transmission yields

$$\chi_{ee}^{xx} = 2j\left(\frac{A_t e^{jk_{x,i}x} - A_i e^{jk_{x,t}x}}{A_i k_{z,i} e^{jk_{x,i}x} + A_t k_{z,t} e^{jk_{x,i}x}}\right), \tag{7.14a}$$

$$\chi_{mm}^{yy} = \frac{2j}{k_0^2}\left(\frac{A_t k_{z,t} e^{jk_{x,i}x} - A_i k_{z,i} e^{jk_{x,t}x}}{A_t e^{jk_{x,i}x} + A_i e^{jk_{x,t}x}}\right). \tag{7.14b}$$

Let us now consider the susceptibilities in (7.12). Consistent with the previous case, we impose $R_x = 0$, while the transmission coefficient may be expressed as

$$T_x = \frac{\Psi_t}{\Psi_i}, \tag{7.15}$$

where $\Psi_i = e^{-jk_{x,i}x}$ and $\Psi_t = e^{-jk_{x,t}x}$ are the phase profiles of the incident and the transmitted waves in the metasurface plane, respectively. Substituting (7.15) into (7.12) yields

$$\chi_{ee}^{xx} = \chi_{mm}^{yy} = \frac{2}{k_0}\tan\left(\frac{k_{x,t} - k_{x,i}}{2}x\right). \tag{7.16}$$

As can be seen by comparing (7.14) and (7.16), the two synthesis methods yield quite different susceptibility results. The main difference is that the susceptibilities in (7.14) are complex, whereas those of (7.16) are purely real. Therefore, according to the discussion in Section 2.5, we expect that the susceptibilities in (7.14) represent gain, loss, or both, whereas those in (7.16) correspond to a gainless and lossless metasurface.

Let us now investigate the efficiency of these two methods. For this purpose, we consider the specifications $\theta_i = 20°$ and $\theta_t = 45°$. Additionally, we also impose conservation of energy, i.e. $A_t = A_i\sqrt{\cos\theta_i/\cos\theta_t}$, which is a generalization of (2.83) for obliquely propagating waves. To understand how the susceptibilities in (7.14) behave in terms of gain and loss, we substitute them into (2.72), which provides the corresponding surface power density in terms of position on the metasurface plane. The resulting spatial evolution of the electric surface power density $\langle I_P \rangle$ and its magnetic counterpart $\langle I_M \rangle$ are plotted in Figure 7.6. These two functions reveal a particularly interesting phenomenon. The synthesized metasurface follows here the specified conservation of energy between the incident and refracted waves by spatially combining electric gain and magnetic loss (and vice versa) so that, overall, gain and loss cancel each other. This results in a metasurface that appears fully efficient to an external observer but that is in fact simultaneously active and lossy.

Let us now consider the other synthesis method and its susceptibilities in (7.16). Since these susceptibilities are purely real, the metasurface is gainless and lossless but that does not necessarily imply that its refraction efficiency, defined as the ratio of the power refracted at θ_t over the power incident at θ_i, is 100%. Indeed, as

Figure 7.6 Metasurface surface power density contributions with $A_i = 1$ along the metasurface plane. Positive values correspond to loss and negative values to gain.

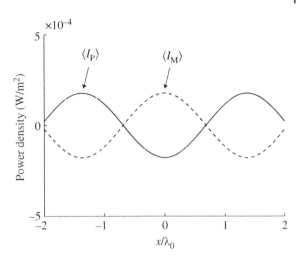

mentioned above, the relations (7.12) are only approximations and do not exactly describe the synthesis problem considered here since they assume normally propagating waves. In fact, a metasurface with the susceptibilities (7.16) cannot perfectly match both incident and transmitted waves. Consequently, it would scatter part of the incident power in undesired directions. The amount of spurious scattering can be assessed by evaluating the mismatch between the incident and transmitted waves. Using the concept of surface wave impedance discussed in Section 7.2, we can define the impedance of the incident and transmitted waves, using (7.3), as

$$Z_i = \eta_0 / \cos\theta_i, \tag{7.17a}$$

$$Z_t = \eta_0 / \cos\theta_t, \tag{7.17b}$$

respectively. The mismatch between the two waves may then be expressed as $(Z_t - Z_i)/(Z_t + Z_i)$ [133], which leads to the metasurface refraction efficiency [147]

$$\zeta = 1 - \left(\frac{Z_t - Z_i}{Z_t + Z_i}\right)^2 = \frac{4\cos\theta_i \cos\theta_t}{(\cos\theta_i + \cos\theta_t)^2}. \tag{7.18}$$

For illustration, we plot (7.18) as a function of θ_i and θ_t in Figure 7.7, which shows that the synthesis method based on (7.12) indeed fails to provide the desired perfect refraction efficiency for waves propagating with relatively large angles as the efficiency quickly drops to zero for grazing angles.

Finally, we use the finite-difference frequency-domain (FDFD) analysis method presented in Section 5.2 to compute the fields scattered by metasurfaces designed with the two methods discussed above for $\theta_i = 20°$ and $\theta_t = 45°$. The simulation results are plotted in Figure 7.8, where a Gaussian beam illumination is used instead of a plane wave for better visualization. We see that the active and lossy

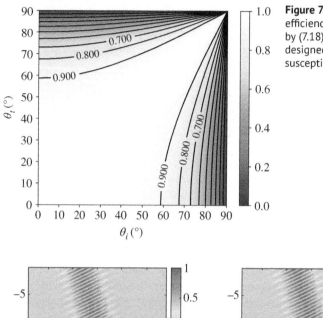

Figure 7.7 Refraction efficiency computed by (7.18) for a metasurface designed with the susceptibilities (7.16).

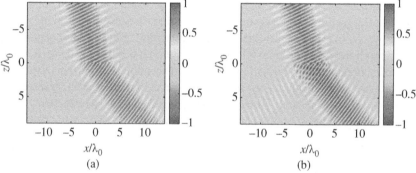

Figure 7.8 Full-wave simulated real part of $H_z \eta_0$ obtained using (a) the first synthesis method, with the susceptibilities (7.14), (b) the second synthesis method with susceptibilities (7.16).

metasurface designed with (7.14) performs the specified refraction transformation without any spurious scattering, as expected from its exact synthesis, whereas the metasurface designed with (7.16) exhibits undesired scattering, due to its approximate synthesis.

7.3.2 Perfect Refraction Using Bianisotropy

We shall now see that it is possible to design a gainless and lossless refractive metasurface that exhibits 100% refraction efficiency for all angles [88]. As mentioned in Section 7.2, when the surface impedance of waves coming from the two sides of the metasurface is different, it is usually better to synthesize the metasurface by separately specifying the direct and reverse transformations.

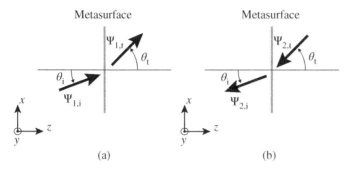

Figure 7.9 Direct and reverse transformations for synthesizing a fully efficient gainless and lossless refractive metasurface.

Since according to (7.17), the impedances of the incident and transmitted waves are not the same, we next synthesize the metasurface according to the two transformations depicted in Figure 7.9, where Figure 7.9b is the reverse transformation of that in Figure 7.9a. In these figures, $\Psi_{1/2,i}$ and $\Psi_{1/2,t}$ represent the incident and transmitted waves of the first and second transformations, respectively. The transverse fields corresponding to the waves of the first transformation in Figure 7.9a are, from (7.13), given by

$$E_{x1,i} = \frac{k_{z,i}}{k_0}e^{-jk_{x,i}x}, \quad E_{x1,t} = A_t\frac{k_{z,t}}{k_0}e^{-jk_{x,t}x}, \tag{7.19a}$$

$$H_{y1,i} = e^{-jk_{x,i}x}/\eta_0, \quad H_{y1,t} = A_t e^{-jk_{x,t}x}/\eta_0, \tag{7.19b}$$

where we have set $A_i = 1$ for normalization and $A_t = A_i\sqrt{\cos\theta_i/\cos\theta_t}$ for energy conservation. The fields corresponding to the second transformation are obtained by simply reversing the sign of the k-vectors in (7.19), i.e., $k_2 = -k_1$, which gives

$$E_{x2,i} = -\frac{k_{z,i}}{k_0}e^{jk_{x,i}x}, \quad E_{x2,t} = -A_t\frac{k_{z,t}}{k_0}e^{jk_{x,t}x}, \tag{7.20a}$$

$$H_{y2,i} = e^{jk_{x,i}x}/\eta_0, \quad H_{y2,t} = A_t e^{jk_{x,t}x}/\eta_0. \tag{7.20b}$$

The synthesis of the metasurface is now performed by substituting (7.19) and (7.20) into (7.7), applying (4.4) and solving for the susceptibilities, which yields

$$\chi_{ee}^{xx} = \frac{4\sin(\alpha x)}{\beta\cos(\alpha x) + \sqrt{\beta^2 - \gamma^2}}, \tag{7.21a}$$

$$\chi_{mm}^{yy} = \frac{\beta^2 - \gamma^2}{4k_0^2}\frac{4\sin(\alpha x)}{\beta\cos(\alpha x) + \sqrt{\beta^2 - \gamma^2}}, \tag{7.21b}$$

$$\chi_{em}^{xy} = -\chi_{me}^{yx} = \frac{2j}{k_0}\frac{\gamma\cos(\alpha x)}{\beta\cos(\alpha x) + \sqrt{\beta^2 - \gamma^2}}, \tag{7.21c}$$

where $\beta = k_{z,i} + k_{z,t}$, $\gamma = k_{z,i} - k_{z,t}$, and $\alpha = k_{x,t} - k_{x,i}$. It can be easily verified from Table 2.2 that the bianisotropic susceptibilities (7.21) correspond to a reciprocal, gainless, and lossless refractive metasurface.

Let us now verify by full-wave simulation the gainless, lossless, and fully efficient scattering properties of the refractive metasurfaces corresponding to (7.21). For this purpose, we synthesize two metasurfaces: one with a large refraction angle, namely $(\theta_i, \theta_t) = (0°, -70°)$, and one that performs negative refraction with $(\theta_i, \theta_t) = (20°, -28°)$, both at around 10 GHz. We consider TM-polarized waves and scattering particles based on the three-metallic-layers unit cell described in Section 6.3 but with x-oriented dog-bone scatterers instead of cross potent for simplicity. The synthesis procedure is the same for both metasurfaces. It consists in (i) substituting the specified angles into (7.21), (ii) discretizing the resulting periodic susceptibility functions into different unit cells, (iii) transforming these discrete susceptibility values into scattering parameters using (4.16), and (iv) optimizing the dimensions of the dog-bone scatterers using full-wave simulations until each unit cell exhibits the required scattering behavior. The final dimensions and material parameters for the two metasurfaces are given in [88].

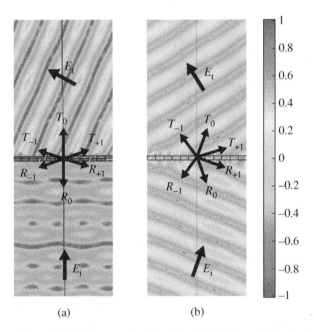

(a) (b)

Figure 7.10 Full-wave simulated electric field magnitude for the two refractive metasurfaces [88]. The arrows indicate the direction of wave propagation and of the propagating diffraction orders. (a) Metasurface with $(\theta_i, \theta_t) = (0°, -70°)$ at 10.5 GHz. (b) Metasurface with $(\theta_i, \theta_t) = (20°, -28°)$ at 10 GHz.

The full-wave simulation results are plotted in Figure 7.10, where we have indicated the propagation direction of the incident and transmitted waves as well as the metasurface diffraction orders.

The diffracted powers of the diffraction orders are plotted in Figure 7.11 for the two metasurfaces. As can be seen, both metasurfaces behave as expected and thus refract the incident wave in the desired direction with very high efficiency. The only reason why the refraction efficiency does not reach 100% is due to the inevitable intrinsic loss of the material used to implement the scattering particles. Interestingly, the maximum refraction efficiency, for the refraction operation in Figure 7.10a, reaches 83%. This is higher than the maximum refraction efficiency

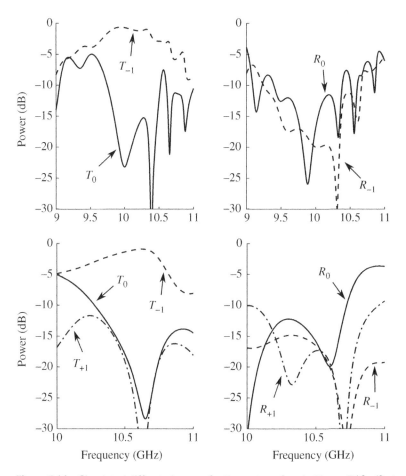

Figure 7.11 Simulated diffracted power for the metasurface in Figure 7.10a (first row) and the that in Figure 7.10b (second row). Note that the T_{+1} and R_{+1} orders are not visible in the first row plots as their magnitudes are lower than −80 dB [88].

of a *non-bianisotropic* metasurface (see Figure 7.7), which does not exceed 75% for the same set of incident and transmission angles. This proves that a bianisotropic metasurface is the best design choice for a refraction operation even taking into account material loss.

Finally, note that the synthesis procedure discussed in this section only applies to transformations for which the normal power flow through the metasurface can be equalized between the waves at the two sides of it and at any spatial position on the metasurface plane [42]. This is obviously the case for the refraction operation addressed above, since we have been able to enforce energy conservation in (7.19) and (7.20). However, this would be impossible to achieve in the case of, for instance, a metasurface lens, as that designed in Sections 5.1 and 5.2. Indeed, in this case, the field at the transmit side of the metasurface in Figure 5.1 decays exponentially along $|x|$, while the field at the incident side is a plane whose amplitude is constant along x. Such a transformation could not be achieved with a passive and lossless metasurface with 100% efficiency due to the impossibility of matching the two waves together. Some of the power must then inevitably be scattered into undesired directions, thus decreasing the metasurface efficiency.

8

Conclusions

Within a decade of spectacular development, metasurfaces have become a novel paradigm of modern science and technology. This book has presented a unified and comprehensive perspective of this new field, based on the powerful framework of bianisotropic surface susceptibilities combined with generalized sheet boundary conditions, and including the latest most significant advances of the field in technology and applications. Moreover, it has presented a number of aspects that extend the frontiers of fundamental physical concepts, such as birefringence, anisotropy, bianisotropy, spatial dispersion, multipolar resonances, Floquet–Bloch theory, electromagnetic scattering, nonlinearity engineering, and surface and leaky wave propagation.

We believe that the content of this book will represent a lasting foundation for the field of metasurfaces. However, at the time of this writing, hundreds of new papers are being published every month on the topic, and it is naturally impossible to predict the evolution of such a massive research effort. Nevertheless, state-of-the art metasurfaces still suffer from some outstanding issues that should clearly spur intense research developments in the forthcoming years. Among these issues are the following: (i) the existing susceptibility-to-scattering particle shape mapping techniques are still too rudimentary for efficient design; (ii) the current metasurfaces feature strong temporal dispersion, which has been little investigated (characterized, mitigated, or leveraged); (iii) some scattering particles exhibit higher order (than bianisotropic) spatial dispersion that is not yet accounted for in the current susceptibility-GSTC model. Fortunately, several strategies have already been suggested to mitigate these issues. We shall conclude this book by briefly discussing these strategies.

- The current procedures to determine the optimal shape for metasurface scattering particles are still time-consuming, inefficient, and limited, due to as the simplistic nature of their modeling approaches (see Chapter 6). The development of better models relating the shapes of the particles to their

Electromagnetic Metasurfaces: Theory and Applications, First Edition.
Karim Achouri and Christophe Caloz.
© 2021 The Institute of Electrical and Electronics Engineers, Inc.
Published 2021 by John Wiley & Sons, Inc.

scattering responses, for instance using symmetry considerations extending those presented in Section 6.4, and effectively accounting for the coupling between adjacent scattering particles, especially in spatially varying metasurfaces, are greatly needed for faster and more efficient metasurface design. This represents a challenge that may greatly benefit from the help of mathematical tools such as group theory and topology, and computational tools such as artificial intelligence and machine learning. The ultimate goal of such developments would be the automation of the metasurface synthesis procedure, from user specifications all the way to the final metasurface with optimal scattering particles.

- As we have seen on many occasions throughout the book, metasurfaces are strongly temporally dispersive, since they are based on resonant scattering particles. Therefore, they exhibit a limited bandwidth of operation. There are cases where a limited bandwidth is not an issue, for instance, in monochromatic applications. A limited bandwidth may even be beneficial, for instance for the implementation of filtering metasurfaces. However, there are many applications that require broadband capabilities, such as for instance white light lenses and thermal radiation emitters. In such cases, temporal dispersion leads to spurious effects such as chromatic aberrations or undesired magnitude modulation and must be properly addressed. Possible solutions include multilayer configurations and highly coupled resonances, which may either increase the overall bandwidth or provide acceptable multiwavelength operation capabilities.

- Spatial dispersion, briefly described as the angular dependence of the structure in Section 2.3, does not play a significant role in the case of deeply subwavelength scattering particles, whose electromagnetic behavior is dominated by dipolar responses. In contrast, it becomes significant in the case of electrically larger particles such as the dielectric resonators discussed in Section 6.2.1.2. For this kind of structures, the excitation of high-order multipoles, especially electric and magnetic quadrupoles and octopoles, spatial dispersion may even play a dominant role that could extend the range of metasurface capabilities beyond those offered in the homogeneous regime. Such effects will need to be included to the current GSTC model for an accurate and efficient description of the angular response of metasurfaces.

9

Appendix

9.1 Approximation of Average Fields at an Interface

Consider a zero-thickness interface in the xy-plane at $z = 0$ that separates region 1 from region 2. We next assume that this interface consists of a sheet of effective impedances, polarizabilities, or susceptibilities, as discussed in Section 3.3. The presence of this interface induces a discontinuity of the fields, as shown in Figure 9.1, where only the electric fields in both regions are represented for convenience.

We will now derive an expression for the average fields at the position of the interface. For this purpose, we note that the total electric field may be written everywhere in space as

$$\mathbf{E}_{\text{tot}}(z) = \mathbf{E}_1(z)U(-z) + \mathbf{E}_2(z)U(z), \tag{9.1}$$

where $U(z)$ is the unit step (or Heaviside) function. The average electric field at the interface follows from (9.1) as

$$\mathbf{E}_{\text{av}}(0) = \lim_{z \to 0} \mathbf{E}_{\text{tot}}(z) = \lim_{z \to 0} \left[\mathbf{E}_1(z)U(-z) + \mathbf{E}_2(z)U(z) \right]. \tag{9.2}$$

Unfortunately, $U(0)$ is undefined. However, we may assume that the transition from $\mathbf{E}_1$ to $\mathbf{E}_2$ across the interface occurs in a smooth and continuous fashion since the system considered is a physical one and is thus anyway continuous at the microscopic scale. In that case, the Heaviside function may be advantageously expressed in terms of a continuous function such as, for instance, the logistic function, i.e.

$$U(z) = \lim_{p \to \infty} \frac{1}{2} \left[1 + \tanh(pz) \right], \tag{9.3}$$

Electromagnetic Metasurfaces: Theory and Applications, First Edition.
Karim Achouri and Christophe Caloz.
© 2021 The Institute of Electrical and Electronics Engineers, Inc.
Published 2021 by John Wiley & Sons, Inc.

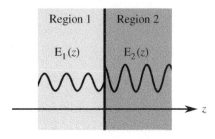

Figure 9.1 Two regions separated by a zero-thickness interface inducing a discontinuity of the fields.

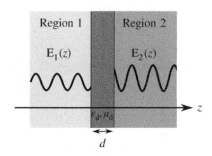

Figure 9.2 Two regions separated by a slab of thickness d and parameters ϵ_d and μ_d.

where p is a parameter related to the sharpness of the transition of the function around $z = 0$. For a large value of p, we have that $U(0) \approx 1/2$. We may therefore approximate the average electric and magnetic fields at the interface as

$$\mathbf{E}_{av} \approx \frac{1}{2}(\mathbf{E}_1 + \mathbf{E}_2), \tag{9.4a}$$

$$\mathbf{H}_{av} \approx \frac{1}{2}(\mathbf{H}_1 + \mathbf{H}_2). \tag{9.4b}$$

Since Eq. (9.4a) are only approximations, we now set to find a more accurate expression for the fields at the interface. For this purpose, consider a uniform slab of thickness d and parameters ϵ_d and μ_d, which lies in the xy-plane between $z = 0$ and $z = d$, as shown in Figure 9.2. We shall now derive an expression for the average fields inside this slab following [164].

The tangential electric field inside the slab as function of z may be expressed as

$$\mathbf{E}_t(z) = \mathbf{A}e^{-j\beta z} + \mathbf{B}e^{j\beta z}, \tag{9.5}$$

where $\mathbf{A}$ and $\mathbf{B}$ are the complex amplitudes of the forward and backward waves propagating within the slab with $\beta = \sqrt{k^2 - k_t^2}$, where k_t is the transverse component of the propagation vector in the slab. Applying the boundary conditions at both interfaces of the slab implies that

$$\mathbf{E}_t(0) = \mathbf{E}_{t1} \quad \text{and} \quad \mathbf{E}_t(d) = \mathbf{E}_{t2}, \tag{9.6}$$

where $\mathbf{E}_{t1}$ and $\mathbf{E}_{t2}$ are the tangential components of the fields just before and after the slab, respectively. The amplitudes $\mathbf{A}$ and $\mathbf{B}$ in (9.5) may be obtained by substituting (9.6) into (9.5), which gives

$$\mathbf{A} = \frac{\mathbf{E}_{t2} - \mathbf{E}_{t1}e^{j\beta d}}{e^{-j\beta d} - e^{j\beta d}} \quad \text{and} \quad \mathbf{B} = -\frac{\mathbf{E}_{t2} - \mathbf{E}_{t1}e^{-j\beta d}}{e^{-j\beta d} - e^{j\beta d}}. \tag{9.7}$$

To obtain the expression of the average field inside the slab, we integrate (9.5) over its thickness. This leads, with (9.7), to

$$\mathbf{E}_{t,av} = \frac{1}{d}\int_0^d \mathbf{E}_t(z)dz = \frac{(\mathbf{E}_{t1} + \mathbf{E}_{t2})}{\beta d} \tan\left(\frac{\beta d}{2}\right). \tag{9.8}$$

In the limit $d \to 0$, this expression reduces to

$$\lim_{d\to 0} \mathbf{E}_{t,av} = \frac{1}{2}(\mathbf{E}_{t1} + \mathbf{E}_{t2}). \tag{9.9}$$

Thus, the average field on a zero-thickness sheet is simply the arithmetic average of the fields on both sides of it. Similar considerations obviously apply to the magnetic field. In the case where the thickness of the sheet is not zero but very small compared to the wavelength, i.e. $\beta d \ll 1$, then we may use the Taylor expansion of (9.8) beyond the first order, i.e.

$$\mathbf{E}_{t,av} = \frac{1}{2}(\mathbf{E}_{t1} + \mathbf{E}_{t2})\left[1 + \frac{1}{12}(\beta d)^2 + \frac{1}{120}(\beta d)^4 + \cdots\right]. \tag{9.10}$$

9.2 Fields Radiated by a Sheet of Dipole Moments

Consider an infinite sheet lying in the xy-plane at $z = 0$ consisting of a subwavelength periodic array of electric and magnetic dipole moments. We are interested in finding the fields scattered by this sheet. For this purpose, we transform the dipole moments into the equivalent surface current densities, $\mathbf{J} = j\omega\mathbf{p}/S$ and $\mathbf{K} = j\omega\mathbf{m}/S$, respectively, where S is the area of the unit cell. This means that the sheet essentially supports a uniform surface current distributions whose scattered electromagnetic fields may be found, using the vector potentials $\mathbf{A}$ and $\mathbf{F}$, as [54]

$$\mathbf{E} = -j\omega\mathbf{A} + \frac{1}{j\omega\mu\epsilon}\nabla(\nabla\cdot\mathbf{A}) - \frac{1}{\epsilon}\nabla\times\mathbf{F}, \tag{9.11a}$$

$$\mathbf{H} = \frac{1}{\mu}\nabla\times\mathbf{A} - j\omega\mathbf{F} + \frac{1}{j\omega\mu\epsilon}\nabla(\nabla\cdot\mathbf{F}), \tag{9.11b}$$

where the vector potentials are related to the current densities as

$$\mathbf{A}(\mathbf{r}) = \frac{\mu}{4\pi}\int_{V_0} \mathbf{J}(\mathbf{r}')\frac{e^{-jkR}}{R}dV', \tag{9.12a}$$

$$\mathbf{F}(\mathbf{r}) = \frac{\epsilon}{4\pi}\int_{V_0} \mathbf{K}(\mathbf{r}')\frac{e^{-jkR}}{R}dV'. \tag{9.12b}$$

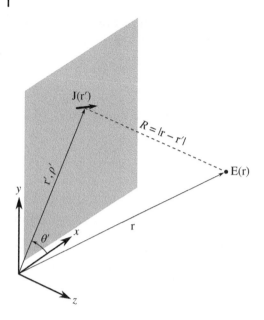

Figure 9.3 System of coordinate. The currents that lie on the xy-plane at $z = 0$ are associated with prime coordinates, while the point of observation where the fields are computed is associated with non-prime coordinates.

For illustration, we now evaluate (9.12a) in cylindrical coordinates for a uniform surface current sheet with $\mathbf{J}(\mathbf{r}') = \mathbf{J}_s \delta(z')$, where $\mathbf{J}_s$ is the sheet current assumed to be purely tangential, i.e. $\mathbf{J}_s \cdot \hat{\mathbf{z}} = 0$. The corresponding system of coordinates is depicted in Figure 9.3. To compute the vector potential (9.12), it is sufficient to consider that $\mathbf{r} = \hat{\mathbf{z}}z$ due to the symmetry of the problem, which leads to $R = \sqrt{\rho'^2 + z^2}$.

Substituting these definitions of the current and R into (9.12a) yields

$$\mathbf{A}(\mathbf{r}) = \frac{\mu}{4\pi} \mathbf{J}_s \int_0^{2\pi} \int_0^\infty \frac{e^{-jk\sqrt{\rho'^2+z^2}}}{\sqrt{\rho'^2 + z^2}} \rho' d\rho' d\theta' = \frac{\mu}{2} \mathbf{J}_s \int_0^\infty \frac{e^{-jk\sqrt{\rho'^2+z^2}}}{\sqrt{\rho'^2 + z^2}} \rho' d\rho'.$$

(9.13)

The last integral in (9.13) can be computed by considering the change of variable $\phi' = k\sqrt{\rho'^2 + z^2}$, which implies that $\rho' d\rho' = \sqrt{\rho'^2 + z^2} d\phi'/k$. This also changes the interval of the integral from $[0, \infty[$ to $[k|z|, \infty[$, hence transforming (9.13) into

$$\mathbf{A}(\mathbf{r}) = \frac{\mu}{2k} \mathbf{J}_s \int_{k|z|}^\infty e^{-j\phi'} d\phi' = \frac{j\mu}{2k} \mathbf{J}_s \left[e^{-j\phi'}\right]_{k|z|}^\infty = -\frac{j\mu}{2k} \mathbf{J}_s e^{-jk|z|} + C,$$

(9.14)

where the constant C is proportional to $e^{-j\infty}$, which is a bounded value that may be ignored in further steps since it vanishes due to the derivatives in (9.11). Following the exact same procedure for the vector potential $\mathbf{F}$, we obtain

$$\mathbf{F}(\mathbf{r}) = -\frac{j\epsilon}{2k} \mathbf{K}_s e^{-jk|z|}.$$

(9.15)

The fields scattered by the sheet may now be found by substituting (9.14) and (9.15) into (9.11). Note that $\nabla \cdot \mathbf{A} = \nabla \cdot \mathbf{F} = 0$ in (9.11) since the currents are assumed to be spatially uniform. The scattered fields are then

$$\mathbf{E}_s = -\frac{\eta}{2}\mathbf{J}_s e^{-jk|z|} + \frac{j}{2k}\nabla \times \left(\mathbf{K}_s e^{-jk|z|}\right), \tag{9.16a}$$

$$\mathbf{H}_s = -\frac{j}{2k}\nabla \times \left(\mathbf{J}_s e^{-jk|z|}\right) - \frac{1}{2\eta}\mathbf{K}_s e^{-jk|z|}, \tag{9.16b}$$

from which we may express the fields right after, $\mathbf{\Psi}_s^+(z = 0^+)$ where $|z| = z$, and right before, $\mathbf{\Psi}_s^-(z = 0^-)$ where $|z| = -z$, the sheet. They respectively read

$$\mathbf{E}_s^+ = -\frac{1}{2}\left(\eta\mathbf{J}_s - \hat{\mathbf{z}} \times \mathbf{K}_s\right) \quad \text{and} \quad \mathbf{H}_s^+ = -\frac{1}{2}\left(\frac{1}{\eta}\mathbf{K}_s + \hat{\mathbf{z}} \times \mathbf{J}_s\right), \tag{9.17a}$$

$$\mathbf{E}_s^- = -\frac{1}{2}\left(\eta\mathbf{J}_s + \hat{\mathbf{z}} \times \mathbf{K}_s\right) \quad \text{and} \quad \mathbf{H}_s^- = -\frac{1}{2}\left(\frac{1}{\eta}\mathbf{K}_s - \hat{\mathbf{z}} \times \mathbf{J}_s\right). \tag{9.17b}$$

These fields may also be expressed in terms of the dipole moments as

$$\mathbf{E}_s^+ = -\frac{j\omega}{2S}\left(\eta\mathbf{p} - \hat{\mathbf{z}} \times \mathbf{m}\right) \quad \text{and} \quad \mathbf{H}_s^+ = -\frac{j\omega}{2S}\left(\frac{1}{\eta}\mathbf{m} + \hat{\mathbf{z}} \times \mathbf{p}\right), \tag{9.18a}$$

$$\mathbf{E}_s^- = -\frac{j\omega}{2S}\left(\eta\mathbf{p} + \hat{\mathbf{z}} \times \mathbf{m}\right) \quad \text{and} \quad \mathbf{H}_s^- = -\frac{j\omega}{2S}\left(\frac{1}{\eta}\mathbf{m} - \hat{\mathbf{z}} \times \mathbf{p}\right). \tag{9.18b}$$

9.3 Relations Between Susceptibilities and Polarizabilities

Here, we provide the expressions relating the susceptibilities, used throughout this work, and the polarizabilities used in the associated synthesis technique discussed in Section 3.3.2. The susceptibilities and the polarizabilities may be related to each other through the expressions of the microscopic electric and magnetic dipole moments, $\mathbf{p}$ and $\mathbf{m}$, and the surface polarization densities, $\mathbf{P}$ and $\mathbf{M}$, as $\mathbf{P} = \mathbf{p}/S$ and $\mathbf{M} = \mathbf{m}/S$, where S is the unit cell area. The first step is to express the difference and the average of the fields on both sides of the metasurface using the formalism in [13, 114]. From (9.18) and assuming vacuum as the background medium, the backward radiated fields from the metasurface-induced dipole moments are

$$\mathbf{E}_p^- = -\frac{j\omega}{2S}\eta_0\mathbf{p}, \quad \mathbf{E}_m^- = \frac{j\omega}{2S}\mathbf{n} \times \mathbf{m}, \tag{9.19}$$

where the normal vector $\mathbf{n} = -\hat{\mathbf{z}}$. Similarly, the forward radiated fields are

$$\mathbf{E}_p^+ = -\frac{j\omega}{2S}\eta_0\mathbf{p}, \quad \mathbf{E}_m^+ = -\frac{j\omega}{2S}\mathbf{n} \times \mathbf{m}. \tag{9.20}$$

Using these expressions, we may define the electric fields on both sides of the metasurface as

$$\mathbf{E}|_{z=0^-} = \mathbf{E}_{inc} + \mathbf{E}_p^- + \mathbf{E}_m^-, \tag{9.21a}$$

$$\mathbf{E}|_{z=0^+} = \mathbf{E}_{inc} + \mathbf{E}_p^+ + \mathbf{E}_m^+. \tag{9.21b}$$

Noting that, for plane waves propagating along z, the magnetic field may be expressed as $\mathbf{H} = -\mathbf{n} \times \mathbf{E}/\eta_0$, which leads to

$$\mathbf{H}|_{z=0^-} = -\mathbf{n} \times \frac{\mathbf{E}_{inc}}{\eta_0} + \mathbf{n} \times \frac{\mathbf{E}_p^-}{\eta_0} + \mathbf{n} \times \frac{\mathbf{E}_m^-}{\eta_0}, \tag{9.22a}$$

$$\mathbf{H}|_{z=0^+} = -\mathbf{n} \times \frac{\mathbf{E}_{inc}}{\eta_0} - \mathbf{n} \times \frac{\mathbf{E}_p^+}{\eta_0} - \mathbf{n} \times \frac{\mathbf{E}_m^+}{\eta_0}. \tag{9.22b}$$

Finally, upon insertion of (9.19) and (9.20) into (9.21) and (9.22), the average fields are

$$\mathbf{E}_{av} = \frac{1}{2}\left(\mathbf{E}|_{z=0^-} + \mathbf{E}|_{z=0^+}\right) = \mathbf{E}_{inc} - \frac{j\omega\eta_0}{2S}\mathbf{p}, \tag{9.23a}$$

$$\mathbf{H}_{av} = \frac{1}{2}\left(\mathbf{H}|_{z=0^-} + \mathbf{H}|_{z=0^+}\right) = \mathbf{H}_{inc} - \frac{j\omega}{2S\eta_0}\mathbf{m}. \tag{9.23b}$$

Now that the average fields have been obtained as functions of the induced dipole moments, we may use the microscopic constitutive relations to relate the polarization densities to the polarizabilities. The microscopic constitutive relations, expressed in terms of the incident electromagnetic fields and the effective polarizabilities, are given in (3.33) as

$$\mathbf{p} = \bar{\bar{\alpha}}_{ee} \cdot \mathbf{E}_{inc} + \bar{\bar{\alpha}}_{em} \cdot \mathbf{H}_{inc}, \tag{9.24a}$$

$$\mathbf{m} = \bar{\bar{\alpha}}_{me} \cdot \mathbf{E}_{inc} + \bar{\bar{\alpha}}_{mm} \cdot \mathbf{H}_{inc}. \tag{9.24b}$$

Substituting (9.23) into (9.24) and making use of the fact that $\mathbf{P} = \mathbf{p}/S$ and $\mathbf{M} = \mathbf{m}/S$, leads to

$$\mathbf{P} \cdot S = \bar{\bar{\alpha}}_{ee} \cdot \left(\mathbf{E}_{av} + \frac{j\omega\eta_0}{2}\mathbf{P}\right) + \bar{\bar{\alpha}}_{em} \cdot \left(\mathbf{H}_{av} + \frac{j\omega}{2\eta_0}\mathbf{M}\right), \tag{9.25a}$$

$$\mathbf{M} \cdot S = \bar{\bar{\alpha}}_{me} \cdot \left(\mathbf{E}_{av} + \frac{j\omega\eta_0}{2}\mathbf{P}\right) + \bar{\bar{\alpha}}_{mm} \cdot \left(\mathbf{H}_{av} + \frac{j\omega}{2\eta_0}\mathbf{M}\right). \tag{9.25b}$$

This linear system of equations may be solved so as to express $\mathbf{P}$ and $\mathbf{M}$ as functions of the effective polarizabilities as

$$\mathbf{P} = \bar{\bar{C}}_p^{-1} \cdot \left[\bar{\bar{\alpha}}_{ee} \cdot \mathbf{E}_{av} + \bar{\bar{\alpha}}_{em} \cdot \mathbf{H}_{av} + \bar{\bar{\alpha}}_{em}\frac{j\omega}{2\eta_0} \cdot \left(S\bar{\bar{I}}_t - \bar{\bar{\alpha}}_{mm}\frac{j\omega}{2\eta_0}\right)^{-1} \right. $$
$$\left. \cdot \left(\bar{\bar{\alpha}}_{me} \cdot \mathbf{E}_{av} + \bar{\bar{\alpha}}_{mm} \cdot \mathbf{H}_{av}\right)\right], \tag{9.26a}$$

$$\mathbf{M} = \overline{\overline{C}}_m^{-1} \cdot \left[\overline{\overline{\tilde{\alpha}}}_{me} \cdot \mathbf{E}_{av} + \overline{\overline{\tilde{\alpha}}}_{mm} \cdot \mathbf{H}_{av} + \overline{\overline{\tilde{\alpha}}}_{me} \frac{j\omega\eta_0}{2} \cdot \left(S\overline{\overline{I}}_t - \overline{\overline{\tilde{\alpha}}}_{ee} \frac{j\omega\eta_0}{2} \right)^{-1} \right.$$
$$\left. \cdot \left(\overline{\overline{\tilde{\alpha}}}_{ee} \cdot \mathbf{E}_{av} + \overline{\overline{\tilde{\alpha}}}_{em} \cdot \mathbf{H}_{av} \right) \right], \tag{9.26b}$$

where $\overline{\overline{I}}_t$ is the tangential unit dyadic and the tensors $\overline{\overline{C}}_p$ and $\overline{\overline{C}}_m$ read

$$\overline{\overline{C}}_p = S\overline{\overline{I}}_t - \overline{\overline{\tilde{\alpha}}}_{ee} \frac{j\omega\eta_0}{2} + \frac{\omega^2}{4} \overline{\overline{\tilde{\alpha}}}_{em} \cdot \left(S\overline{\overline{I}}_t - \overline{\overline{\tilde{\alpha}}}_{mm} \frac{j\omega}{2\eta_0} \right)^{-1} \cdot \overline{\overline{\tilde{\alpha}}}_{me}, \tag{9.27a}$$

$$\overline{\overline{C}}_m = S\overline{\overline{I}}_t - \overline{\overline{\tilde{\alpha}}}_{mm} \frac{j\omega}{2\eta_0} + \frac{\omega^2}{4} \overline{\overline{\tilde{\alpha}}}_{me} \cdot \left(S\overline{\overline{I}}_t - \overline{\overline{\tilde{\alpha}}}_{ee} \frac{j\omega\eta_0}{2} \right)^{-1} \cdot \overline{\overline{\tilde{\alpha}}}_{em}. \tag{9.27b}$$

In (9.26), the polarization densities, $\mathbf{P}$ and $\mathbf{M}$, have been expressed in terms of the microscopic effective polarizabilities. They can now be related to the macroscopic susceptibilities via the constitutive relations used throughout this work and which are given in (2.2) as

$$\mathbf{P} = \epsilon_0 \overline{\overline{\chi}}_{ee} \cdot \mathbf{E}_{av} + \overline{\overline{\chi}}_{em} \frac{1}{c_0} \cdot \mathbf{H}_{av}, \tag{9.28a}$$

$$\mathbf{M} = \overline{\overline{\chi}}_{mm} \cdot \mathbf{H}_{av} + \overline{\overline{\chi}}_{me} \frac{1}{\eta_0} \cdot \mathbf{E}_{av}. \tag{9.28b}$$

Finally, the susceptibilities can be expressed as functions of the polarizabilities by inserting (9.26) into (9.28) and associating the terms that multiply the average electric and magnetic fields in both equations. This leads to the following relations

$$\overline{\overline{\chi}}_{ee} = \frac{1}{\epsilon_0} \overline{\overline{C}}_p^{-1} \cdot \left[\overline{\overline{\tilde{\alpha}}}_{ee} + \frac{j\omega}{2\eta_0} \overline{\overline{\tilde{\alpha}}}_{em} \cdot \left(S\overline{\overline{I}}_t - \overline{\overline{\tilde{\alpha}}}_{mm} \frac{j\omega}{2\eta_0} \right)^{-1} \cdot \overline{\overline{\tilde{\alpha}}}_{me} \right], \tag{9.29a}$$

$$\overline{\overline{\chi}}_{em} = c_0 \overline{\overline{C}}_p^{-1} \cdot \left[\overline{\overline{\tilde{\alpha}}}_{em} + \frac{j\omega}{2\eta_0} \overline{\overline{\tilde{\alpha}}}_{em} \cdot \left(S\overline{\overline{I}}_t - \overline{\overline{\tilde{\alpha}}}_{mm} \frac{j\omega}{2\eta_0} \right)^{-1} \cdot \overline{\overline{\tilde{\alpha}}}_{mm} \right], \tag{9.29b}$$

$$\overline{\overline{\chi}}_{me} = \eta_0 \overline{\overline{C}}_m^{-1} \cdot \left[\overline{\overline{\tilde{\alpha}}}_{me} + \frac{j\omega\eta_0}{2} \overline{\overline{\tilde{\alpha}}}_{me} \cdot \left(S\overline{\overline{I}}_t - \overline{\overline{\tilde{\alpha}}}_{ee} \frac{j\omega\eta_0}{2} \right)^{-1} \cdot \overline{\overline{\tilde{\alpha}}}_{ee} \right], \tag{9.29c}$$

$$\overline{\overline{\chi}}_{mm} = \overline{\overline{C}}_m^{-1} \cdot \left[\overline{\overline{\tilde{\alpha}}}_{mm} + \frac{j\omega\eta_0}{2} \overline{\overline{\tilde{\alpha}}}_{me} \cdot \left(S\overline{\overline{I}}_t - \overline{\overline{\tilde{\alpha}}}_{ee} \frac{j\omega\eta_0}{2} \right)^{-1} \cdot \overline{\overline{\tilde{\alpha}}}_{em} \right]. \tag{9.29d}$$

We see that the polarizabilities are related to the susceptibilities in a non-straightforward fashion. As can be seen, for instance, the purely electric susceptibility, $\overline{\overline{\chi}}_{ee}$, is related to all bianisotropic polarizabilities except in the case where either of the magnetoelectric coupling tensors, $\overline{\overline{\tilde{\alpha}}}_{em}$ and $\overline{\overline{\tilde{\alpha}}}_{me}$, are zero.

References

1 Ahmed H. Abdelrahman, Atef Z. Elsherbeni, and Fan Yang. Transmitarray antenna design using cross-slot elements with no dielectric substrate. *IEEE Antennas Wirel. Propag. Lett.*, 13:177–180, 2014.

2 Karim Achouri and Christophe Caloz. Design, concepts, and applications of electromagnetic metasurfaces. *Nanophotonics*, 7(6):1095–1116, June 2018.

3 Karim Achouri and Olivier J. F. Martin. Surface-wave dispersion retrieval method and synthesis technique for bianisotropic metasurfaces. *Phys. Rev. B*, 99:155140, Apr. 2019.

4 Karim Achouri and Olivier J. F. Martin. Angular Scattering Properties of Metasurfaces. *IEEE Trans. Antennas Propag.*, 8:1–1, 2019.

5 Karim Achouri and Olivier J. F. Martin. Angular scattering properties of metasurfaces. *IEEE Trans. Antennas Propag.*, 68(1):432–442, Jan. 2020.

6 Karim Achouri, Mohamed A. Salem, and Christophe Caloz. General metasurface synthesis based on susceptibility tensors. *IEEE Trans. Antennas Propag.*, 63(7):2977–2991, Jul. 2015.

7 Karim Achouri, Guillaume Lavigne, and Christophe Caloz. Comparison of two synthesis methods for birefringent metasurfaces. *J. Appl. Phys.*, 120(23):235305, 2016.

8 Karim Achouri, Gabriel D. Bernasconi, Jérémy Butet, and Olivier J. F. Martin. Homogenization and scattering analysis of second-harmonic generation in nonlinear metasurfaces. *IEEE Trans. Antennas Propag.*, 66 (11):6061–6075, 2018.

9 Vladimir M. Agranovich and Vitaly Ginzburg. *Crystal optics with spatial dispersion, and excitons*, volume 42. Springer Science & Business Media, 2013.

10 M. Albooyeh, D.-H. Kwon, F. Capolino, and S. A. Tretyakov. Equivalent realizations of reciprocal metasurfaces: role of tangential and normal polarization. *Phys. Rev. B*, 95(11):115435, Mar. 2017.

Electromagnetic Metasurfaces: Theory and Applications, First Edition.
Karim Achouri and Christophe Caloz.
© 2021 The Institute of Electrical and Electronics Engineers, Inc.
Published 2021 by John Wiley & Sons, Inc.

11 Amir Arbabi, Yu Horie, Mahmood Bagheri, and Andrei Faraon. Dielectric metasurfaces for complete control of phase and polarization with sub-wavelength spatial resolution and high transmission. *Nat. Nanotechnol.*, 10:937–9432015.

12 V. S. Asadchy, Y. Ra'di, J. Vehmas, and S. A. Tretyakov. Functional metamirrors using bianisotropic elements. *Phys. Rev. Lett.*, 114:095503, Mar. 2015.

13 Viktar Asadchy. *Spatially dispersive metasurfaces*. PhD thesis, Aalto University, 2017.

14 Viktar S. Asadchy, Igar A. Faniayeu, Younes Ra'di, and Sergei A. Tretyakov. Determining polarizability tensors for an arbitrary small electromagnetic scatterer. *Photon. Nanostruct. Fundam. Appl.*, 12(4): 298–304, 2014.

15 Viktor S. Asadchy and Igor A. Fanyaev. Simulation of the electromagnetic properties of helices with optimal shape, which provides radiation of a circularly polarized wave. *J. Adv. Res. Phys.*, 2(1):011107, 2011.

16 A. A. Barybin. Modal expansions and orthogonal complements in the theory of complex media waveguide excitation by external sources for isotropic, anisotropic, and bianisotropic media. *Prog. Electromagn. Res.*, 19:241–300, 1998.

17 D. Berry, R. Malech, and W. Kennedy. The reflectarray antenna. *IEEE Trans. Antennas Propag.*, 11(6):645–651, Nov. 1963.

18 Michael V. Berry. The adiabatic phase and pancharatnam's phase for polarized light. *J. Mod. Opt.*, 34(11):1401–1407, 1987.

19 Borui Bian, Shaobin Liu, Shenyun Wang, Xiangkun Kong, Haifeng Zhang, Ben Ma, and Huan Yang. Novel triple-band polarization-insensitive wide-angle ultra-thin microwave metamaterial absorber. *J. Appl. Phys.*, 114(19):194511, 2013.

20 Jagadis Chunder Bose. On the rotation of plane of polarisation of electric waves by a twisted structure. *Proc. R. Soc. A*, 63(389–400):146–152, 1898.

21 Robert W. Boyd. Nonlinear optics. *Handbook of laser technology and applications (three-volume set)*, pages 161–183. Taylor & Francis, 2003.

22 John Brown. Artificial dielectrics having refractive indices less than unity. *Proc. IEE-Part IV Inst. Monogr.*, 100(5):51–62, 1953.

23 L. Van Buskirk and C. Hendrix. The zone plate as a radio-frequency focusing element. *IRE Trans. Antennas Propag.*, 9(3):319–320, May 1961.

24 Christophe Caloz and Zoé-Lise Deck-Léger. Spacetime metamaterials. *arXiv preprint arXiv:1905.00560*, 2019.

25 Christophe Caloz and Tatsuo Itoh. *Electromagnetic metamaterials: transmission line theory and microwave applications*. Wiley, 2005.

26 Christophe Caloz and Ari Sihvola. Electromagnetic chirality. *arXiv preprint arXiv:1903.09087*, 2019.

27 Christophe Caloz and Ari Sihvola. Electromagnetic chirality, part 1: the microscopic perspective [electromagnetic perspectives]. *IEEE Antennas Propag. Mag.*, 62(1):58–71, Feb. 2020.

28 Christophe Caloz, Andrea Alù, Sergei Tretyakov, Dimitrios Sounas, Karim Achouri, and Zoé-Lise Deck-Léger. Electromagnetic nonreciprocity. *Phys. Rev. Appl.*, 10:047001, Oct. 2018.

29 Filippo Capolino. *Theory and phenomena of metamaterials.* CRC Press, 2009.

30 Nima Chamanara, Karim Achouri, and Christophe Caloz. Efficient analysis of metasurfaces in terms of spectral-domain GSTC integral equations. *IEEE Trans. Antennas Propag.*, 65(10):5340–5347, Oct. 2017.

31 Robert E. Collin. *Field theory of guided waves.* Wiley-IEEE Press, 1991.

32 John F. Cornwell. *Group theory in physics: an introduction*, volume 1. Academic Press, 1997.

33 Haixia Da, Qiaoliang Bao, Roozbeh Sanaei, Jinghua Teng, Kian Ping Loh, Francisco J. Garcia-Vidal, and Cheng-Wei Qiu. Monolayer graphene photonic metastructures: giant Faraday rotation and nearly perfect transmission. *Phys. Rev. B*, 88:205405, Nov. 2013.

34 S. Datthanasombat, A. Prata, L. R. Arnaro, J. A. Harrell, S. Spitz, and J. Perret. Layered lens antennas. *IEEE Antennas and Propagation Society International Symposium. 2001 Digest. Held in conjunction with: USNC/URSI National Radio Science Meeting (Cat. No.01CH37229)*, volume 2, pages 777–780, Jul. 2001.

35 Manuel Decker, Stefan Linden, and Martin Wegener. Coupling effects in low-symmetry planar split-ring resonator arrays. *Opt. Lett.*, 34(10): 1579–1581, May 2009.

36 A. Degiron and D. R. Smith. One-way glass for microwaves using nonreciprocal metamaterials. *Phys. Rev. E*, 89:053203, May 2014.

37 Mojtaba Dehmollaian, Yousef Vahabzadeh, Karim Achouri, and Christophe Caloz. Metasurface modeling by a thin slab. *arXiv preprint arXiv:2005.00901*, 2020.

38 N. Destouches, A. V. Tishchenko, J. C. Pommier, S. Reynaud, O. Parriaux, S. Tonchev, and M. Abdou Ahmed. 99% efficiency measured in the -1st order of a resonant grating. *Opt. Express*, 13(9):3230–3235, May 2005.

39 Andrea Di Falco, Yang Zhao, and Andrea Alù. Optical metasurfaces with robust angular response on flexible substrates. *Appl. Phys. Lett.*, 99(16): 163110, 2011.

40 Furkan Dincer, Oguzhan Akgol, Muharrem Karaaslan, Emin Unal, and Cumali Sabah. Polarization angle independent perfect metamaterial absorbers for solar cell applications in the microwave, infrared, and visible regime. *Prog. Electromagn. Res.*, 144:93–101, 2014.

41 Nader Engheta and Richard W. Ziolkowski. *Metamaterials: physics and engineering explorations*. Wiley, 2006.

42 A. Epstein and G. V. Eleftheriades. Arbitrary power-conserving field transformations with passive lossless omega-type bianisotropic metasurfaces. *IEEE Trans. Antennas Propag.*, 64(9):3880–3895, Sep. 2016.

43 B. H. Fong, J. S. Colburn, J. J. Ottusch, J. L. Visher, and D. F. Sievenpiper. Scalar and tensor holographic artificial impedance surfaces. *IEEE Trans. Antennas Propag.*, 58(10):3212–3221, Oct. 2010.

44 N. Gagnon, A. Petosa, and D.A. McNamara. Thin microwave quasi-transparent phase-shifting surface (PSS). *IEEE Trans. Antennas Propag.*, 58(4):1193–1201, Apr. 2010.

45 Ju Gao, Kuang Zhang, Guohui Yang, and Qun Wu. A novel four-face polarization twister based on three-dimensional magnetic toroidal dipoles. *IEEE Trans. Mag.*, 50(1):1–4, Jan. 2014.

46 P. Genevet and F. Capasso. Breakthroughs in photonics 2013: flat optics: wavefronts control with Huygens' interfaces. *Photon. J. IEEE*, 6(2):1–4, Apr. 2014.

47 Dylan Germain, Divitha Seetharamdoo, Shah Nawaz Burokur, and André de Lustrac. Phase-compensated metasurface for a conformal microwave antenna. *Appl. Phys. Lett.*, 103(12):124102, 2013.

48 Stanislav B. Glybovski, Sergei A. Tretyakov, Pavel A. Belov, Yuri S. Kivshar, and Constantin R. Simovski. Metasurfaces: from microwaves to visible. *Phys. Rep.*, 634:1–72, 2016.

49 Joseph W. Goodman et al. *Introduction to Fourier optics*, volume 2. McGraw-Hill, New York, 1968.

50 J. A. Gordon, C. L. Holloway, and A. Dienstfrey. A physical explanation of angle-independent reflection and transmission properties of metafilms/metasurfaces. *IEEE Antennas Wirel. Propag. Lett.*, 8:1127–1130, 2009.

51 Anthony Grbic, Lei Jiang, and Roberto Merlin. Near-field plates: subdiffraction focusing with patterned surfaces. *Science*, 320(5875): 511–513, 2008.

52 Sybren Ruurd de Groot and Leendert Gerrit Suttorp. *Foundations of electrodynamics*. North-Holland, 1972.

53 Roger F. Harrington. *Field computation by moment methods*. Wiley-IEEE Press, 1993.

54 Roger F. Harrington. *Time-harmonic electromagnetic fields*. Wiley-Interscience, 2001.

55 C. L. Holloway. Sub-wavelength resonators: on the use of metafilms to overcome the $\lambda/2$ size limit. *IET Microwaves Antennas Propag.*, 2:120–129(9), Mar. 2008.

56 C. L. Holloway, M. A. Mohamed, Edward F. Kuester, and A. Dienstfrey. Reflection and transmission properties of a metafilm: with an application to a controllable surface composed of resonant particles. *IEEE Trans. Electromagn. Compat.*, 47(4):853–865, Nov. 2005.

57 C. L. Holloway, Andrew Dienstfrey, Edward F. Kuester, John F. O'Hara, Abul K. Azad, and Antoinette J. Taylor. A discussion on the interpretation and characterization of metafilms/metasurfaces: the two-dimensional equivalent of metamaterials. *Metamaterials*, 3(2):100–112, Oct. 2009.

58 C. L. Holloway, Edward F. Kuester, J. A. Gordon, J. O'Hara, J. Booth, and D. R. Smith. An overview of the theory and applications of metasurfaces: the two-dimensional equivalents of metamaterials. *IEEE Antennas Propag. Mag.*, 54(2):10–35, Apr. 2012.

59 Ben Yu-Kuang Hu. Kramers–Kronig in two lines. *Am. J. Phys.*, 57(9): 821, 1989.

60 Cheng Huang, Xiaoliang Ma, Mingbo Pu, Guangwei Yi, Yanqin Wang, and Xiangang Luo. Dual-band 90° polarization rotator using twisted split ring resonators array. *Opt. Commun.*, 291:345–348, 2013.

61 John Huang and José A. Encinar. *Reflectarray antenna*. Wiley Online Library, 2007.

62 John Hunt, Tom Driscoll, Alex Mrozack, Guy Lipworth, Matthew Reynolds, David Brady, and David R. Smith. Metamaterial apertures for computational imaging. *Science*, 339(6117):310–313, 2013.

63 John Hunt, Jonah Gollub, Tom Driscoll, Guy Lipworth, Alex Mrozack, Matthew S. Reynolds, David J. Brady, and David R. Smith. Metamaterial microwave holographic imaging system. *J. Opt. Soc. Am. A*, 31(10): 2109–2119, Oct. 2014.

64 M. Mithat Idemen. *Discontinuities in the electromagnetic field*. Wiley, 2011.

65 M. F. Imani and A. Grbic. Planar near-field plates. *IEEE Trans. Antennas Propag.*, 61(11):5425–5434, Nov. 2013.

66 Mohammadreza F. Imani and Anthony Grbic. An analytical investigation of near-field plates. *Metamaterials*, 4(2):104–111, 2010.

67 Akira Ishimaru. *Electromagnetic wave propagation, radiation, and scattering*. Prentice-Hall, 1991.

68 Tatsuo Itoh. *Numerical techniques for microwave and millimeter-wave passive structures*. Wiley-Interscience, 1989.

69 Shang-Chi Jiang, Xiang Xiong, Yuan-Sheng Hu, Yu-Hui Hu, Guo-Bin Ma, Ru-Wen Peng, Cheng Sun, and Mu Wang. Controlling the polarization state of light with a dispersion-free metastructure. *Phys. Rev. X*, 4:021026, May 2014.

70 Jian-Ming Jin. *Theory and computation of electromagnetic fields*. Wiley, 2011.

71 John D. Joannopoulos, Steven G. Johnson, Joshua N. Winn, and Robert D. Meade. *Photonic crystals: molding the flow of light*. Princeton University Press, 2011.

72 Peter B. Johnson and R. W. Christy. Optical constants of the noble metals. *Phys. Rev. B*, 6(12):4370, 1972.

73 R. Clark Jones. A new calculus for the treatment of optical systemsi. description and discussion of the calculus. *J. Opt. Soc. Am.*, 31(7): 488–493, Jul. 1941.

74 M. A. Joyal and J. J. Laurin. Analysis and design of thin circular polarizers based on meander lines. *IEEE Trans. Antennas Propag.*, 60(6):3007–3011, June 2012.

75 P. J. Kahrilas. Hapdar 8212: an operational phased array radar. *Proc. IEEE*, 56(11):1967–1975, Nov. 1968.

76 Ebrahim Karimi, Sebastian A. Schulz, Israel De Leon, Hammam Qassim, Jeremy Upham, and Robert W. Boyd. Generating optical orbital angular momentum at visible wavelengths using a plasmonic metasurface. *Light Sci. Appl.*, 3(5):e167, 2014.

77 G. Kenanakis, R. Zhao, A. Stavrinidis, G. Konstantinidis, N. Katsarakis, M. Kafesaki, C. M. Soukoulis, and E. N. Economou. Flexible chiral metamaterials in the terahertz regime: a comparative study of various designs. *Opt. Mater. Express*, 2(12):1702–1712, Dec. 2012.

78 Milton Kerker. *The scattering of light and other electromagnetic radiation: physical chemistry: a series of monographs*, volume 16. Academic Press, 2013.

79 Winston E. Kock. Metallic delay lenses. *Bell Labs Techn. J.*, 27(1):58–82, 1948.

80 T. Kodera, D. L. Sounas, and C. Caloz. Artificial Faraday rotation using a ring metamaterial structure without static magnetic field. *Appl. Phys. Lett.*, 99(3):031114, Jul. 2011.

81 J. A. Kong. *Electromagnetic wave theory*. A Wiley-Interscience publication. Wiley, 1986.

82 Alexander Krasnok, Sergey Makarov, Mikhail Petrov, Roman Savelev, Pavel Belov, and Yuri Kivshar. Towards all-dielectric metamaterials and nanophotonics. *Proc. SPIE*, 9502:950203–950203–17, 2015.

83 Alexander Krasnok, Mykhailo Tymchenko, and Andrea Alu. Nonlinear metasurfaces: a paradigm shift in nonlinear optics. *Materials Today*, 21 (1):8–21, 2018.

84 Sergey Kruk and Yuri Kivshar. 5-Tailoring transmission and reflection with metasurfaces. *Dielectric metamaterials*, pages 145–174. Woodhead Publishing, 2020.

85 Edward F. Kuester, M. A. Mohamed, M. Piket-May, and C. L. Holloway. Averaged transition conditions for electromagnetic fields at a metafilm. *IEEE Trans. Antennas Propag.*, 51(10):2641–2651, Oct 2003.

86 King-Wai Lam, Sze-Wai Kwok, Yeongming Hwang, and T. K. Lo. Implementation of transmitarray antenna concept by using aperture-coupled microstrip patches. *Proceedings of 1997 Asia-Pacific Microwave Conference*, volume 1, pages 433–436, Dec. 1997.

87 Horace Lamb. On the reflection and transmission of electric waves by a metallic grating. *Proc. Lond. Math. Soc.*, 1(1):523–546, 1897.

88 G. Lavigne, K. Achouri, V. S. Asadchy, S. A. Tretyakov, and C. Caloz. Susceptibility derivation and experimental demonstration of refracting metasurfaces without spurious diffraction. *IEEE Trans. Antennas Propag.*, 66(3):1321–1330, 2018.

89 Jongwon Lee, Mykhailo Tymchenko, Christos Argyropoulos, Pai-Yen Chen, Feng Lu, Frederic Demmerle, Gerhard Boehm, Markus-Christian Amann, Andrea Alù, and Mikhail A. Belkin. Giant nonlinear response from plasmonic metasurfaces coupled to intersubband transitions. *Nature*, 511(7507):65–69, 2014.

90 Quentin Levesque, Mathilde Makhsiyan, Patrick Bouchon, Fabrice Pardo, Julien Jaeck, Nathalie Bardou, Christophe Dupuis, Riad Haïdar, and Jean-Luc Pelouard. Plasmonic planar antenna for wideband and efficient linear polarization conversion. *Appl. Phys. Lett.*, 104(11):111105, 2014.

91 Minhua Li, Linyan Guo, Jianfeng Dong, and Helin Yang. An ultra-thin chiral metamaterial absorber with high selectivity for LCP and RCP waves. *J. Phys. D*, 47(18):185102, 2014.

92 Dianmin Lin, Pengyu Fan, Erez Hasman, and Mark L. Brongersma. Dielectric gradient metasurface optical elements. *Science*, 345(6194): 298–302, 2014.

93 Jiao Lin, Patrice Genevet, Mikhail A. Kats, Nicholas Antoniou, and Federico Capasso. Nanostructured holograms for broadband manipulation of vector beams. *Nano Lett.*, 13(9):4269–4274, 2013.

94 Karl F. Lindman. Über eine durch ein isotropes system von spiralförmigen resonatoren erzeugte rotationspolarisation der elektromagnetischen wellen. *Ann. Phys.*, 368(23):621–644, 1920.

95 Guy Lipworth, Alex Mrozack, John Hunt, Daniel L. Marks, Tom Driscoll, David Brady, and David R. Smith. Metamaterial apertures for coherent computational imaging on the physical layer. *J. Opt. Soc. Am. A*, 30(8): 1603–1612, Aug. 2013.

96 Fu Liu, Shiyi Xiao, A. Sihvola, and Jensen Li. Perfect co-circular polarization reflector: a class of reciprocal perfect conductors with total co-circular polarization reflection. *IEEE Trans. Antennas Propag.*, 62(12): 6274–6281, Dec. 2014.

97 S. Maci, G. Minatti, M. Casaletti, and M. Bosiljevac. Metasurfing: addressing waves on impenetrable metasurfaces. *IEEE Antennas Wirel. Propag. Lett.*, 10:1499–1502, 2011.

98 C. S. Malagisi. Microstrip disc element reflect array. *EASCON'78: Electronics and Aerospace Systems Convention*, volume 1, pages 186–192, 1978.

99 L. Martinez-Lopez, J. Rodriguez-Cuevas, J. I. Martinez-Lopez, and A. E. Martynyuk. A multilayer circular polarizer based on bisected split-ring frequency selective surfaces. *IEEE Antennas Wirel. Propag. Lett.*, 13: 153–156, 2014.

100 James Clerk Maxwell. *A treatise on electricity and magnetism*, volume 1. Clarendon Press, 1881.

101 D. McGrath. Planar three-dimensional constrained lenses. *IEEE Trans. Antennas Propag.*, 34(1):46–50, Jan. 1986.

102 Kevin M. McPeak, Sriharsha V. Jayanti, Stephan J. P. Kress, Stefan Meyer, Stelio Iotti, Aurelio Rossinelli, and David J. Norris. Plasmonic films can easily be better: rules and recipes. *ACS Photon.*, 2(3):326–333, 2015.

103 Christoph Menzel, Carsten Rockstuhl, and Falk Lederer. Advanced jones calculus for the classification of periodic metamaterials. *Phys. Rev. A*, 82: 053811, Nov. 2010.

104 Gustav Mie. Beiträge zur optik trüber medien, speziell kolloidaler metallösungen. *Ann. Phys.*, 330(3):377–445, 1908.

105 Alexander E. Minovich, Andrey E. Miroshnichenko, Anton Y. Bykov, Tatiana V. Murzina, Dragomir N. Neshev, and Yuri S. Kivshar. Functional and nonlinear optical metasurfaces. *Laser Photonics Rev.*, 9(2):195–213, Mar. 2015.

106 Kenro Miyamoto. The phase fresnel lens. *J. Opt. Soc. Am.*, 51(1):17–20, Jan. 1961.

107 J. P. Montgomery. A microstrip reflectarray antenna element. *Annales d'Astrophysique*. Antenna Applications Symposium, University of Illinois, 1978.

108 Francesco Monticone, Nasim Mohammadi Estakhri, and Andrea Alù. Full control of nanoscale optical transmission with a composite metascreen. *Phys. Rev. Lett.*, 110:203903, May 2013.

109 Stefan Mühlig, Christoph Menzel, Carsten Rockstuhl, and Falk Lederer. Multipole analysis of meta-atoms. *Metamaterials*, 5(2–3):64–73, 2011.

110 Ben A. Munk. *Frequency selective surfaces: theory and design*. Wiley, 2000.

111 H. Nematollahi, J. J. Laurin, J. E. Page, and J. A. Encinar. Design of broadband transmitarray unit cells with comparative study of different numbers of layers. *IEEE Trans. Antennas Propag.*, 63(4):1473–1481, Apr. 2015.

112 M. Neviere. Electromagnetic study of transmission gratings. *Appl. Opt.*, 30(31):4540–4547, Nov. 1991.

113 T. Niemi, A. O. Karilainen, and S. A. Tretyakov. Synthesis of polarization transformers. *IEEE Trans. Antennas Propag.*, 61(6):3102–3111, June 2013.

114 Teemu Petteri Niemi. Polarization transformations in bianisotropic arrays. Master's thesis, Aalto University, Finland, 2012.

115 Lukas Novotny and Bert Hecht. *Principles of nano-optics*. Cambridge University Press, 2012.

116 Mark A. Ordal, Robert J. Bell, Ralph W. Alexander, Larry L. Long, and Marvin R. Querry. Optical properties of fourteen metals in the infrared and far infrared: Al, co, cu, au, fe, pb, mo, ni, pd, pt, ag, ti, v, and w. *Appl. Opt.*, 24(24):4493–4499, 1985.

117 J. D. Ortiz, J. D. Baena, V. Losada, F. Medina, and J. L. Araque. Spatial angular filtering by fsss made of chains of interconnected SRRs and CSRRs. *IEEE Microwave Comp. Lett.*, 23(9):477–479, 2013.

118 Robert Paknys. *Applied frequency-domain electromagnetics*. Wiley, Chichester, May 2016.

119 Shivaramakrishnan Pancharatnam. Generalized theory of interference and its applications. *Proceedings of the Indian Academy of Sciences-Section A*, volume 44, pages 398–417. Springer, 1956.

120 J. B. Pendry. Negative refraction makes a perfect lens. *Phys. Rev. Lett.*, 85:3966–3969, Oct. 2000.

121 J. B. Pendry, D. Schurig, and D. R. Smith. Controlling electromagnetic fields. *Science*, 312(5781):1780–1782, 2006.

122 John B. Pendry, Anthony J. Holden, David J. Robbins, W. J. Stewart, et al. Magnetism from conductors and enhanced nonlinear phenomena. *IEEE Trans. Microwave Theory Tech.*, 47(11):2075–2084, 1999.

123 C. Pfeiffer and A. Grbic. Millimeter-wave transmitarrays for wavefront and polarization control. *IEEE Trans. Microwave Theory Tech.*, 61(12): 4407–4417, Dec. 2013.

124 Carl Pfeiffer and Anthony Grbic. Cascaded metasurfaces for complete phase and polarization control. *Appl. Phys. Lett.*, 102(23):231116, 2013.

125 Carl Pfeiffer and Anthony Grbic. Metamaterial Huygens' surfaces: tailoring wave fronts with reflectionless sheets. *Phys. Rev. Lett.*, 110: 197401, May 2013.

126 Carl Pfeiffer and Anthony Grbic. Bianisotropic metasurfaces for optimal polarization control: analysis and synthesis. *Phys. Rev. Appl.*, 2:044011, Oct. 2014.

127 Carl Pfeiffer and Anthony Grbic. Controlling vector bessel beams with metasurfaces. *Phys. Rev. Appl.*, 2:044012, Oct. 2014.

128 Carl Pfeiffer and Anthony Grbic. Generating stable tractor beams with dielectric metasurfaces. *Phys. Rev. B*, 91:115408, Mar. 2015.

129 Carl Pfeiffer, Naresh K. Emani, Amr M. Shaltout, Alexandra Boltasseva, Vladimir M. Shalaev, and Anthony Grbic. Efficient light bending with isotropic metamaterial Huygens' surfaces. *Nano Lett.*, 14(5):2491–2497, 2014.

130 E. Plum, V. A. Fedotov, A. S. Schwanecke, N. I. Zheludev, and Y. Chen. Giant optical gyrotropy due to electromagnetic coupling. *Appl. Phys. Lett.*, 90(22):223113, 2007.

131 Anders Pors, Michael G. Nielsen, and Sergey I. Bozhevolnyi. Analog computing using reflective plasmonic metasurfaces. *Nano Lett.*, 15(1): 791–797, 2015.

132 D. M. Pozar. Flat lens antenna concept using aperture coupled microstrip patches. *Electron. Lett.*, 32(23):2109–2111, Nov. 1996.

133 D. M. Pozar. *Microwave engineering*. Wiley, fourth edition, 2011.

134 Y. Ra'di and S. A. Tretyakov. Angularly-independent Huygens' metasurfaces. *2015 IEEE International Symposium on Antennas and Propagation & USNC/URSI National Radio Science Meeting*, pages 874–875. IEEE, 2015.

135 Y. Ra'di, V. S. Asadchy, and S. A. Tretyakov. Total absorption of electromagnetic waves in ultimately thin layers. *IEEE Trans. Antennas Propag.*, 61(9):4606–4614, Sep. 2013.

136 Y. Ra'di, V. S. Asadchy, and S. A. Tretyakov. Tailoring reflections from thin composite metamirrors. *IEEE Trans. Antennas Propag.*, 62(7): 3749–3760, Jul. 2014.

137 Y. Ra'di, V. S. Asadchy, and S. A. Tretyakov. One-way transparent sheets. *Phys. Rev. B*, 89:075109, Feb. 2014.

138 Heinz Raether. Surface plasmons on smooth surfaces. *Surface plasmons on smooth and rough surfaces and on gratings*, pages 4–39. Springer, 1988.

139 A. K. Rashid, B. Li, and Z. Shen. An overview of three-dimensional frequency-selective structures. *IEEE Antennas Propag. Mag.*, 56(3):43–67, June 2014.

140 Edward J. Rothwell and Michael J. Cloud. *Electromagnetics*. CRC Press, 2008.

141 W. Rotman. Plasma simulation by artificial dielectrics and parallel-plate media. *IRE Trans. Antennas Propag.*, 10(1):82–95, Jan. 1962.

142 Raymond C. Rumpf. Simple implementation of arbitrarily shaped total-field/scattered-field regions in finite-difference frequency-domain. *Prog. Electromagn. Res. B*, 36:221–248, 2012.

143 B. E. A. Saleh and M. C. Teich. *Fundamentals of photonics*. Wiley series in pure and applied optics. Wiley, 2007.

144 Sergei Alexander Schelkunoff and Harald T. Friis. *Antennas: theory and practice*, volume 639. Wiley, New York, 1952.

145 Daniel L. Schodek, Paulo Ferreira, and Michael F. Ashby. *Nanomaterials, nanotechnologies and design: an introduction for engineers and architects*. Butterworth-Heinemann, 2009.

146 D. Schurig, J. J. Mock, B. J. Justice, S. A. Cummer, J. B. Pendry, A. F. Starr, and D. R. Smith. Metamaterial electromagnetic cloak at microwave frequencies. *Science*, 314(5801):977–980, 2006.

147 Michael Selvanayagam and George V. Eleftheriades. Discontinuous electromagnetic fields using orthogonal electric and magnetic currents for wavefront manipulation. *Opt. Express*, 21(12):14409 14429, 2013.

148 A. Serdiukov, Igor Semchenko, S. Tretyakov, and A. Sihvola. *Electromagnetics of bi-anisotropic materials-theory and application*,volume 11. Gordon and Breach Science Publishers, 2001.

149 Claude Elwood Shannon. Communication in the presence of noise. *Proc. IRE*, 37(1):10–21, 1949.

150 Richard A. Shelby, David R. Smith, and Seldon Schultz. Experimental verification of a negative index of refraction. *Science*, 292(5514):77–79, 2001.

151 Yichen Shen, Dexin Ye, Ivan Celanovic, Steven G. Johnson, John D. Joannopoulos, and Marin Soljačić. Optical broadband angular selectivity. *Science*, 343(6178):1499–1501, 2014.

152 Yichen Shen, Dexin Ye, Li Wang, Ivan Celanovic, Lixin Ran, John D. Joannopoulos, and Marin Soljačić. Metamaterial broadband angular selectivity. *Phys. Rev. B*, 90:125422, Sep. 2014.

153 Hongyu Shi, Jianxing Li, Anxue Zhang, Yansheng Jiang, Jiafu Wang, Zhuo Xu, and Song Xia. Gradient metasurface with both polarization-controlled directional surface wave coupling and anomalous reflection. *IEEE Antennas Wirel. Propag. Lett.*, 14:104–107, 2015.

154 Jinhui Shi, Xu Fang, Edward T. F. Rogers, Eric Plum, Kevin F. MacDonald, and Nikolay I. Zheludev. Coherent control of Snell's law at metasurfaces. *Opt. Express*, 22(17):21051–21060, Aug. 2014.

155 A. H. Sihvola, A. J. Viitanen, I. V. Lindell, and S. A. Tretyakov. *Electromagnetic waves in chiral and bi-isotropic media*. The Artech House Antenna Library. Artech House, 1994.

156 Alexandre Silva, Francesco Monticone, Giuseppe Castaldi, Vincenzo Galdi, Andrea Alù, and Nader Engheta. Performing mathematical operations with metamaterials. *Science*, 343(6167):160–163, 2014.

157 Kun Song, Xiaopeng Zhao, Yahong Liu, Quanhong Fu, and Chunrong Luo. A frequency-tunable 90°-polarization rotation device using composite chiral metamaterials. *Appl. Phys. Lett.*, 103(10):101908, 2013.

158 D. L. Sounas and C. Caloz. Electromagnetic non-reciprocity and gyrotropy of graphene. *Appl. Phys. Lett.*, 98(2):021911:1–3, Jan. 2011.

159 D. L. Sounas, T. Kodera, and C. Caloz. Electromagnetic modeling of a magnetless nonreciprocal gyrotropic metasurface. *IEEE Trans. Antennas Propag.*, 61(1):221–231, Jan. 2013.

160 J. Stangel. The thinned lens approach to phased array design. *1965 Antennas and Propagation Society International Symposium*, volume 3, pages 10–13, Aug. 1965.

161 Jingbo Sun, Xi Wang, Tianboyu Xu, Zhaxylyk A. Kudyshev, Alexander N. Cartwright, and Natalia M. Litchinitser. Spinning light on the nanoscale. *Nano Lett.*, 14(5):2726–2729, 2014.

162 Allen Taflove and Susan C. Hagness. *Computational electrodynamics: the finite-difference time-domain method.* Artech House, 2005.

163 Sajjad Taravati and Christophe Caloz. Mixer-duplexer-antenna leaky-wave system based on periodic space-time modulation. *IEEE Trans. Antennas Propag.*, 65(2):442–452, 2016.

164 Sergei Tretyakov. *Analytical modeling in applied electromagnetics.* Artech House, 2003.

165 Sergei A. Tretyakov. A personal view on the origins and developments of the metamaterial concept. *J. Opt.*, 19(1):013002, 2016.

166 Y. Vahabzadeh, N. Chamanara, K. Achouri, and C. Caloz. Computational analysis of metasurfaces. *IEEE J. Multiscale Multiphys. Comput. Techn.*, 3:37–49, 2018.

167 Yousef Vahabzadeh, Karim Achouri, and Christophe Caloz. Simulation of metasurfaces in finite difference techniques. *IEEE Trans. Antennas Propag.*, 64(11):4753–4759, Nov. 2016.

168 C. A. Valagiannopoulos, A. Tukiainen, T. Aho, T. Niemi, M. Guina, S. A. Tretyakov, and C. R. Simovski. Perfect magnetic mirror and simple perfect absorber in the visible spectrum. *Phys. Rev. B*, 91:115305, Mar. 2015.

169 Ventsislav K. Valev, J. J. Baumberg, Ben De Clercq, N. Braz, Xuezhi Zheng, E. J. Osley, Stefaan Vandendriessche, M. Hojeij, C. Blejean, J. Mertens, et al. Nonlinear superchiral meta-surfaces: tuning chirality and disentangling non-reciprocity at the nanoscale. *Adv. Mater.*, 26(24):4074–4081, 2014.

170 V. G. Veselago. The electrodynamics of substances with simultaneously negative values of ϵ and μ. *Physics-Uspekhi*, 10(4):509–514, 1968.

171 Mehdi Veysi, Caner Guclu, and Filippo Capolino. Vortex beams with strong longitudinally polarized magnetic field and their generation by using metasurfaces. *J. Opt. Soc. Am. B*, 32(2):345–354, Feb. 2015.

172 Yongzheng Wen, Wei Ma, Joe Bailey, Guy Matmon, Xiaomei Yu, and Gabriel Aeppli. Planar broadband and high absorption metamaterial using single nested resonator at terahertz frequencies. *Opt. Lett.*, 39(6): 1589–1592, Mar. 2014.

173 Lin Wu, Zhenyu Yang, Yongzhi Cheng, Rongzhou Gong, Ming Zhao, Yu Zheng, Ji'an Duan, and Xiuhua Yuan. Circular polarization converters based on bi-layered asymmetrical split ring metamaterials. *Appl. Phys. A*, 116(2):643–648, 2014.

174 Fan Yang and Yahya Rahmat-Samii. *Electromagnetic band gap structures in antenna engineering.* Cambridge University Press Cambridge, 2009.

175 Quanlong Yang, Jianqiang Gu, Dongyang Wang, Xueqian Zhang, Zhen Tian, Chunmei Ouyang, Ranjan Singh, Jiaguang Han, and Weili Zhang. Efficient flat metasurface lens for terahertz imaging. *Opt. Express*, 22(21): 25931–25939, Oct. 2014.

176 Xunong Yi, Xiaohui Ling, Zhiyou Zhang, Ying Li, Xinxing Zhou, Yachao Liu, Shizhen Chen, Hailu Luo, and Shuangchun Wen. Generation of cylindrical vector vortex beams by two cascaded metasurfaces. *Opt. Express*, 22(14):17207–17215, Jul. 2014.

177 Minyeong Yoo and Sungjoon Lim. Polarization-independent and ultrawideband metamaterial absorber using a hexagonal artificial impedance surface and a resistor–capacitor layer. *IEEE Trans. Antennas Propag.*, 62(5):2652–2658, May 2014.

178 Nanfang Yu and Federico Capasso. Flat optics with designer metasurfaces. *Nature Mater.*, 13(2):139–150, 2014.

179 Nanfang Yu, Patrice Genevet, Mikhail A. Kats, Francesco Aieta, Jean-Philippe Tetienne, Federico Capasso, and Zeno Gaburro. Light propagation with phase discontinuities: generalized laws of reflection and refraction. *Science*, 334(6054):333–337, 2011.

180 Jia-Xin Zhao, Bo-Xun Xiao, Xiao-Jun Huang, and He-Lin Yang. Multiple-band reflective polarization converter based on complementary l-shaped metamaterial. *Microwave Opt. Technol. Lett.*, 57(4):978–983, 2015.

181 Guoxing Zheng, Holger Mühlenbernd, Mitchell Kenney, Guixin Li, Thomas Zentgraf, and Shuang Zhang. Metasurface holograms reaching 80% efficiency. *Nat. Nanotechnol.*, 10(4):308–312, 2015.

182 Yu Zhu, Xiaoyong Hu, Zhen Chai, Hong Yang, and Qihuang Gong. Active control of chirality in nonlinear metamaterials. *Appl. Phys. Lett.*, 106(9): 091109, 2015.

183 R. W. Ziolkowski. Metamaterials: the early years in the usa. *EPJ Appl. Metamater.*, 1:5, 2014.

Index

Electromagnetic Metasurfaces: Theory and Applications, First Edition.
Karim Achouri and Christophe Caloz.
© 2021 The Institute of Electrical and Electronics Engineers, Inc.
Published 2021 by John Wiley & Sons, Inc.

Printed and bound by CPI Group (UK) Ltd, Croydon, CR0 4YY
11/10/2021
03086589-0001